UNDERSTANDING
FLIGHT

UNDERSTANDING FLIGHT

David F. Anderson
Scott Eberhardt

SECOND EDITION

New York   Chicago   San Francisco   Lisbon   London   Madrid
Mexico City   Milan   New Delhi   San Juan   Seoul
Singapore   Sydney   Toronto

Library of Congress Cataloging-in-Publication Data

Anderson, David F.
  Understanding flight / David F. Anderson, Scott Eberhardt.—2nd ed.
      p.   cm.
  Includes index.
  ISBN 978-0-07-162696-5 (alk. paper)
  1. Flight.   2. Aerodynamics.   I. Eberhardt, Scott.   II. Title.
TL570.A69   2010
629.13—dc22   2009022103

8 9 0   DOC   21 20 19 18 17 16

ISBN 978-0-07-162696-5
MHID 0-07-162696-4

Sponsoring Editor: Stephen S. Chapman
Production Supervisor: Pamela A. Pelton
Editing Supervisor: Stephen M. Smith
Project Manager: Patricia Wallenburg, TypeWriting
Copy Editor: James Madru
Proofreader: Paul Tyler
Art Director, Cover: Jeff Weeks
Composition: TypeWriting

Printed and bound by RR Donnelley.

McGraw-Hill books are available at special quantity discounts to use as premiums and sales promotions, or for use in corporate training programs. To contact a representative, please e-mail us at bulksales@mcgraw-hill.com.

This book is printed on acid-free paper.

About the Authors

David F. Anderson is a private pilot and a lifelong flight enthusiast. He has degrees from the University of Washington, Seattle, and a Ph.D. in physics from Columbia University. He has had a 30-year career in high-energy physics at Los Alamos National Laboratory, CERN in Geneva, Switzerland, and the Fermi National Accelerator Laboratory.

Scott Eberhardt is a private pilot who works in high-lift aerodynamics at Boeing Commercial Airplanes Product Development. He has degrees from MIT and a Ph.D. in aeronautics and astronautics from Stanford University. He joined Boeing in 2006 after 20 years on the faculty of the Department of Aeronautics and Astronautics at the University of Washington, Seattle.

CONTENTS

Flight is a relatively simple and widely studied phenomenon. As surprising as it may sound, though, it is more often than not misunderstood. For example, most descriptions of the physics of lift fixate on the shape of the wing (i.e., airfoil) as the key factor in understanding lift. The wings in these descriptions have a bulge on the top so that the air must travel farther over the top than under the wing. Yet we all know that wings fly quite well upside down, where the shape of the wing is inverted. This is demonstrated by the Thunderbirds in Figure I.1, with wings of almost no thickness at all. To cover for this paradox, we sometimes see a description for inverted flight that is different than for normal flight. In reality, the shape of the wing has little to do with how lift is generated, and any description that relies on the shape of the wing is misleading at best. This assertion will be discussed in detail in Chapter 1. It should be noted that the shape of the wing does has everything to do with the efficiency of the wing at cruise speeds and with stall characteristics.

Let us look at three examples of successful wings that clearly violate the descriptions that rely on the shape of the wing as the basis of lift. The first example is a very old design. Figure I.2 shows a photograph of a Curtis 1911 Model D type IV pusher. Clearly, the air travels the same distance over the top and under the bottom of the wing. Yet this airplane flew and was the second airplane purchased by the U.S. Army in 1911.

> During World War II, the length of a belt of 50-caliber machine gun bullets was 27 feet. When a pilot emptied his guns into a single target, he was giving it the "whole nine yards."

Figure I.3 shows the symmetric wing on an aerobatic airplane. The wing is thick and has a great deal of curvature on the top and bottom to give it good stall characteristics and to allow slow flight. The jets in

FIGURE I.1 Two Thunderbirds in flight. (Photograph courtesy of the U.S. Air Force.)

FIGURE I.2 Curtis 1911 Model D type IV pusher. (Photograph courtesy of the U.S. Air Force Museum.)

FIGURE I.3   Symmetric wing on an aerobatic airplane.

Figure I.1 have extremely thin, symmetric wings to allow them to fly fast but at the price of very abrupt stall entry characteristics.

The final example of a wing that violates the idea that lift depends on the shape of the wing is of a very modern wing. Figure I.4 shows the profile of the Whitcomb Supercritical Airfoil [NASA/Langley SC(2)-0714]. This wing is basically flat on top with the curvature on the bottom. Although its shape may seem contrary to the popular view of the shape of wings, this airfoil is the foundation of modern airliner wings.

The emphasis on wing shape in many explanations of lift is based on the *principle of equal transit times*. This assertion mistakenly states the air going around a wing must take the same length of time, whether going over or under, to get to the trailing edge. The argument goes that since the air goes farther over the top of the wing, it has to go faster, and with Bernoulli's principle, we have lift. Knowing that equal transit times is not defensible, the statement is often softened to say that since the air going over the top must go farther, it must go faster. Again,

FIGURE I.4   Whitcomb Supercritical Airfoil.

FIGURE I.5 Airflow around a wing with lift.

however, this is just a variation on the idea of equal transit times. In reality, equal transit times holds only for a wing without lift. Figure I.5 shows a simulation of the airflow around a wing with lift. It is easy to see that the air going over the wing arrives at the trailing edge before the air going under the wing. In fact, the greater the lift, the greater is the difference in arrival times at the trailing edge. Somewhere around World War II this popular assertion began to be taught from grade school to flight training classes. Before this idea permeated flight instruction, the correct idea of a lift as a reaction force was used.

Another erroneous argument that leads one to believe that the shape of the wing is responsible for the generation of lift is the argument that a wing is a half-venturi. The venturi (shown in the top of Figure I.6) works by constricting airflow. As the airflow constricts, it speeds up, much like putting your thumb on the end of a garden hose. Using the Bernoulli principle, the pressure (perpendicular to the flow) in the constriction decreases. This clever device is used to create low pressure to draw fuel into automobile carburetors. The argument for a wing goes like this: Remove the top half of the venturi, and you have a wing, as shown at the bottom of Figure I.6. The problem, as any physics student can tell you, is that there can be no net lift in the picture. If the air enters horizontally and leaves horizontally, how can there be a vertical force? This will be discussed in Chapter 1.

> The air behind the wing is going almost straight down when seen from the ground.

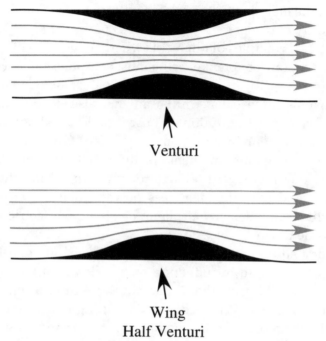

Venturi

Wing
Half Venturi

FIGURE I.6   A venturi and a half-venturi.

We were motivated by these misleading and incorrect descriptions and others to write this book. Starting with lift as a reaction force, a consistent and physically correct description of lift is presented. While the foundation of this description was used commonly over a half-century ago, we have expanded on this simple description to describe many other aspects of flight.

The objective of this book is to present a clear, intuitive description of the phenomenon of flight and of aeronautics without complicated mathematics. This work will be presented on two levels. The bulk of the material will be addressed to the general reader. Here, a minimum of experience will be assumed. At times, it will be desirable to make clarifying comments by insertion of a short topic that may be somewhat removed from the main train of thought. These insertions will be printed on a colored background. These insertions may be skipped over without any loss of continuity or under-

> Unlike previous versions, the tail-plane structure of the Boeing 747-400ER (Extended Range) houses extra fuel.

standing of the main text, although we will try to keep them of interest to the majority of readers.

Chapter 1, "Principles of Flight," is where we get into lift and flight. We believe that this chapter gives the most complete and correct physical description of lift to date. Like many before us, we describe lift using Newton's three laws. Unlike anyone else, to our knowledge, though, we take this description and use it to derive almost all aspects of flight. It allows one to intuitively understand aspects of flight that often are only explained mathematically. It will become clear to the reader why one increases the angle of the wing when the airplane slows down and why lift takes less power when the airplane goes faster. It will be obvious why airplanes can have symmetric wings and can fly upside down.

In the first edition of this book, there was a chapter entitled "Basic Concepts" that was an introduction to a basic set of terms and concepts of aircraft. This gave the reader and the authors a common set of tools with which to begin the discussion of flight and aeronautics. It was found to be too much information all at once and an unnecessarily complicated beginning to an otherwise readable book. We have made this chapter Appendix A in this edition. If the reader is a novice in the field of flight, he or she should read it through quickly and then use it as a reference while reading the book.

> A Cessna 172 at cruise is diverting about five times its own weight in air per second to produce lift.

Two chapters have been added to this edition. The first is a chapter on helicopters and autogyros. We were unable to find a complete and readable discussion of these topics under one cover. Works were either too mathematical and too detailed or very incomplete. Such topics as the physical description of a helicopter's power curve were never discussed in physical terms.

The final chapter is a short discussion of airplane structures. The purpose of this chapter is to give the reader a brief introduction into how airplanes are constructed.

In the end, this book is a complete course in the principles of aeronautics, presented in straightforward, physical terms. We believe that the information will be accessible to nearly all who read it.

UNDERSTANDING
FLIGHT

Principles of Flight

Physical Description of Lift

A jet engine and a propeller produce thrust by blowing air back. A helicopter's rotor produces lift by blowing air down, as can be seen in Figure 1.1, where the downwash of a helicopter hovering over the water is clearly visible. In the same way, a wing produces lift by diverting air down. A jet engine, a propeller, a helicopter's rotor, and a wing all work by the same physics: Air is accelerated in the direction opposite the desired force.

This chapter introduces a physical description of lift. It is based primarily on Newton's three laws. This description is useful for understanding intuitively many phenomena associated with flight that one is not able to understand with other descriptions. This approach allows one to understand in a very clear way how lift changes with such variables as speed, density, load, angle of attack, and wing area. It is valid in low-speed flight as well as supersonic flight. This physical description of lift is also of great use to the pilot who desires an intuitive understanding of the behavior and limitations of his or her airplane. With the knowledge provided in this book, it will be easy to understand why the angle of attack must increase with decreasing speed, why the published maneuvering speed

The characteristic of fluids to have zero velocity at the surface of an object explains why one is not able to hose dust off a car.

FIGURE 1.1   A helicopter pushes air down. (Photograph courtesy of the U.S. Air Force.)

(maximum speed in turbulent air) for an airplane decreases with decreasing load, and why power must be increased for low-speed flight.

Lift is a reaction force. That is, wings develop lift by diverting air down. Since we know that a propeller produces thrust by blowing air back and that a helicopter develops lift by blowing air down, the concept of a wing diverting air down to produce lift should not be difficult to accept. After all, propellers and rotors are simply rotating wings.

One should be careful not to form the mental image of the air striking the bottom of the wing and being deflected down to produce lift. This is a fairly common misconception that also was held by Sir Isaac Newton himself. Since Newton was not familiar with the details of airflow over a wing, he thought that the air was diverted down by its impact with the bottom of a bird's wings. It is true that there can be some lift owing to the diversion of air by the bottom of the wing, but most of the lift is due to the action over the top of the wing. As we will see later, the low pressure that is formed above the wing accelerates the air down.

Newton's Three Laws

The most powerful tools for understanding flight are Newton's three laws of motion. They are simple to understand and universal in application. They apply to the flight of the lowly mosquito and the motion of the galaxies. We will start with a statement of Newton's first law: *A body at rest will remain at rest, and a body in motion will continue in straight-line motion unless acted on by an external applied force.*

In the context of flight, this means that if a mass or blob of air is initially motionless and starts to move, there has been some force acting on it. Likewise, if a flow of air bends, such as over a wing, there also must be a force acting on it. In the context of a continuum such as air, the force expresses itself as a difference in pressure.

> The lift of a wing is proportional to the angle of attack. This is true for all wings, from a modern jet to a barn door.

Going out of order, Newton's third law can be stated: *For every action there is an equal and opposite reaction.*

This is fairly straightforward. When one sits in a chair, you put a force on the chair, and the chair puts an equal and opposite force on you. The force you put on the chair is the action, whereas the force the chair puts on you is the reaction. That is, the chair is reacting to the force you are putting on it. Another example is seen in the case of a bending flow of air over a wing. The bending of the air requires a force from Newton's first law. By Newton's third law, the air must be putting an equal and opposite force on whatever is bending it, in this case the wing. When the air bends down, there must be a downward force on it, and there must be an equal upward force on the wing by Newton's third law. The bending of the air is the action, whereas the lift on the wing is the reaction.

Newton's second law is a little more difficult to understand but also more useful in understanding many phenomena associated with flight. The most common form of the second law, which students are taught in early physics courses, is

$$F = ma$$

or force equals mass times acceleration.

The law in this form gives the force necessary to accelerate an object of a certain mass. For a description of the movement of air, we use an alternative form of this law that can be applied to a jet engine, a rocket, or the lift on a wing. The alternate form of Newton's second law for a rocket can be stated: *The force (or thrust) of a rocket is equal to the amount of gas expelled per time times the velocity of that gas.*

Newton's second law tells us how much thrust is produced by the engine of a rocket. The amount of gas expelled per time might be in units such as pound mass per second (lbm/s) or kilograms per second (kg/s). The velocity of that gas might be in units such as feet per second (ft/s) or meters per second (m/s). To double the thrust, one must double the amount of gas expelled per second, double the velocity of the gas, or a combination of the two.

Let us now look at the airflow around a wing with Newton's laws in mind. Figure 1.2 shows the airflow around a wing as many of us have been shown at one time or another. Notice that the air approaches the wing, splits, and re-forms behind the wing going in the initial direction. This wing has no lift. There is no net *action* on the air, and thus there is no lift, or *reaction* on the wing. If the wing has no net effect on the air, the air cannot have any net effect on the wing. Now look at another picture of air flowing around a wing (Figure 1.3). The air splits around the wing and leaves the wing at a slight downward angle. This downward-traveling air is the *downwash* and is the action that creates lift as its reaction. In this figure, there has been a net change in the air after passing over the wing. Thus there is a force acting on the air and a reaction force acting on the wing. There is lift.

If one were to sum up how a wing generates lift in one sentence, it would be that the wing produces lift by diverting air down. This statement should be as easy to understand as saying that a propeller produces thrust by pushing air back.

FIGURE 1.2   Based on Newton's laws, this wing has no lift.

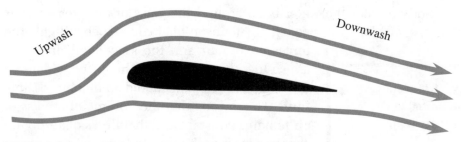

FIGURE 1.3   The airflow around a real wing with lift.

Air Bending Over a Wing

As always, simple statements result in more questions. One natural question is, Why does the air bend around the wing? This question is probably the most challenging question in understanding flight, and it is one of the key concepts.

Let us start by first looking at a simple demonstration. Run a small stream of water from a faucet, and bring a horizontal water glass over to it until it just touches the water, as in Figure 1.4. As in the figure, the water will wrap partway around the glass. From Newton's first law, we know that for the flow of water to bend, there must be a force on it. The force is in the direction of the bend. From Newton's third law, we know that there must be an equal and opposite force acting on the glass. The stream of water puts a force on the glass that tries to pull it into the stream, not push it away, as one might first expect.

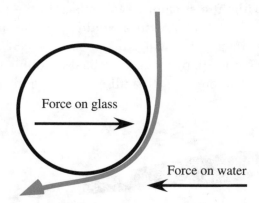

FIGURE 1.4   Water wrapping around a glass.

So why does the water bend around the glass or air over a wing? First, consider low-speed flight (subsonic). In low-speed flight, the forces on the air and the associated pressures are so low that not only is the air considered a fluid but it is also considered an incompressible fluid. This means that the volume of a mass of air remains constant and that flows of air do not separate from each other to form voids (gaps).

> One out of every 11 airplanes registered in the United States flies to Oshkosh every year.

A second point to understand is that streamlines communicate with each other. A streamline, in steady-state flight, can be looked at as the path of a particle in the moving air. It is the path a small, light object would take in the airflow over the wing. The communication between streamlines is expressed as pressure and viscosity. Pressure is the force per area that the air exerts on the neighboring streamline. Viscosity in a gas or liquid corresponds to friction between solids.

Think of two adjacent streamlines with different speeds. Since these streamlines have different velocities, forces between them try to speed up the slower streamline and slow down the faster streamline. The speed of air at the surface of the wing is exactly zero with respect to the surface of the wing. This is an expression of viscosity. The speed of the air increases with distance from the wing, as shown in Figure 1.5. Now imagine that the first non-zero-velocity streamline just grazes the high point of the top of the wing. If it were to go straight back initially and not follow the wing, there would be a volume of zero-velocity air between it and the wing. Forces would strip this air away from the wing, and without a streamline to replace it, the pressure would lower. This lowering of the pressure would bend the streamline until it followed the surface of the wing.

The next streamline above would be bent to follow the first by the same process, and so on. The streamlines increase in speed with dis-

FIGURE 1.5 The speed variations of a fluid near an object.

tance from a wing for a short distance. This is on the order of 6 in (15 cm) at the trailing edge of the wing of an Airbus A380. This region of rapidly changing air speed is called the *boundary layer*. If the boundary layer is not turbulent, the flow is said to be *laminar*.

Thus the streamlines are bent by a lowering of the pressure. This is why the air is bent by the top of the wing and why the pressure above the wing is lowered. This lowered pressure decreases with distance above the wing but is the basis of the lift on a wing. The lowered pressure propagates out at the speed of sound, causing a great deal of air to bend around the wing.

Two streamlines communicate on a molecular scale. This is expressed as the pressure and viscosity of the air. Without viscosity, there would be no communication between streamlines and no boundary layer. Often, calculations of lift are made in the limit of zero viscosity. In these cases, viscosity is reintroduced implicitly with the *Kutta-Joukowski condition*, which requires that the air come smoothly off at the trailing edge of the wing. Also, the calculations require that the air follows the surface of the wing, which is another introduction of the effects of viscosity. One result of the near elimination of viscosity from the calculations is that no boundary layer is calculated.

It should be noted that the speed of the uniform flow over the top of the wing is faster than the *free-stream velocity*, which is the velocity of the undisturbed air some distance from the wing. The bending of the air causes a reduction in pressure above the wing. This reduction in

The Wright brothers did not fly from October 16, 1905, to May 6, 1908, to protect their pending patent.

pressure causes an acceleration of the air. It is often taught that the acceleration of the air causes a reduction in pressure. In fact, it is the reduction of pressure that accelerates the air, in agreement with Newton's first law.

Let us look at the air bending around the wing in Figure 1.6. To bend the air requires a force. As indicated by the colored arrows, the direction of the force on the air is perpendicular to the bend in the air. The magnitude of the force is proportional to the tightness of the bend. The tighter the air bends, the greater is the force on it. The forces on the wing, as shown by the black arrows in the figure, have the same magnitude as the forces on the air, but in the opposite direction. These

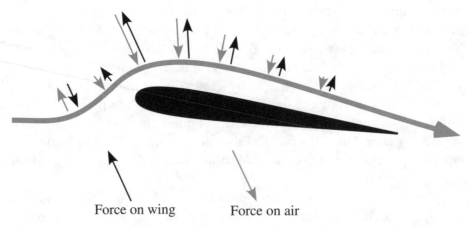

Force on wing   Force on air

FIGURE 1.6 Forces on the air and the corresponding reaction forces on the wing.

forces, working through pressure, represent the mechanism in which the force is transferred to the wing.

Look again at Figure 1.6 while paying attention to the black arrows representing the forces on the wing. There are two points to notice. The first is that most of the lift is on the forward part of the wing. In fact, half the total lift on a wing at subsonic speeds typically is produced in the first one-fourth of the chord length. The *chord* is the distance from the leading edge to the trailing edge of the wing. The second thing to notice is that the arrows on the leading part of the wing are tilted forward. Thus the force of lift is pulling the wing along as well as lifting it. This would be nice if it were the entire story. Unfortunately, the horizontal forces on the trailing part of the wing compensate the horizontal forces on the leading part of the wing.

> Because of the oil embargo, the price of 1000 gallons of Jet A1 fuel went from $100 in 1970 to $1100 in 1980.

We now have the tools to understand why a wing has lift. In brief, the air bends around the wing, producing downwash. Newton's first law says that the bending of the air requires a force on the air, and Newton's third law says that there is an equal and opposite force on the wing. That is a description of lift. The pressure difference across the wing is the mechanism by which lift is transferred to the wing owing to the bending of the air.

Downwash

In the simplest form, lift is generated by the wing diverting air down, creating the downwash. Figure 1.7 is a good example of the effect of downwash behind an airplane. In the picture, the jet has flown above the fog, not through it. The hole caused by the descending air is clearly visible. As we will see, it is the adjustment of the magnitude of the downwash that allows the wing to adjust for varying loads and speeds.

From Newton's second law, one can state the relationship between the lift on a wing and its downwash: *The lift of a wing is proportional to the amount of air diverted per time times the vertical velocity of that air.*

FIGURE 1.7   A jet flying over fog demonstrates downwash. (Photograph by Paul Bowen; courtesy of Cessna Aircraft Co.)

Similar to a rocket, the lift of a wing can be increased by increasing the amount of air diverted, the vertical velocity of that air, or a combination of the two. The concept of the vertical velocity of the downwash may seem a little foreign at first. We are all used to thinking of the airflow across a wing as seen by the pilot or as seen in a wind tunnel. In this *rest frame*, the wing is stationary, and the air is moving. However, what does flight look like in a rest frame where the air is initially standing still and the wing is moving? Picture yourself on top of a mountain. Now suppose that just as a passing airplane is opposite you, you could take a picture of all the velocities of the air. What would you see? You might be surprised.

The first thing you would notice is that the air behind the wing is going almost straight down when seen from the ground. Because of friction with the wing, the air has, in fact, a slight forward direction. So how do we reconcile the two rest frames: the wing stationary and the air moving and the wing moving and the air stationary? Take a look at Figure 1.8a. The arrow labeled "Speed" is the direction and speed of the wing through the air. The arrow labeled "Downwash" is the direction and speed of the air as seen by the pilot or the engineer in the wind tunnel. The colored arrow labeled "V_v" is the vertical velocity of the air seen by the observer on the mountain. V_v is the vertical velocity of the downwash and represents the component that produces lift. In this figure, the letter α indicates the angle of the wing's downwash with respect to the relative wind, which is related to the *effective angle of attack* of the wing. Effective angle of attack will be discussed in the next section.

> The engines on a Boeing 777 have a diameter that is within inches of the fuselage diameter of a Boeing 737.

The plausibility of the statement that the air comes off the wing vertically when the wing flies by is fairly easy to demonstrate. Turn on a small household fan, and examine the tightness of the column of air. If the air were coming off the trailing edges of the fan blades (which are legitimate wings) other than perpendicular to the direction of the blade's motion, the air would form a cone rather than a tight column. This also can be seen in a picture of a helicopter hovering above water (see Figure 1.1). The pattern on the water is the same size as the rotor blades. It is fortunate that nature works this way. If the air behind the propeller of an airplane came off as a cone rather than a column, pro-

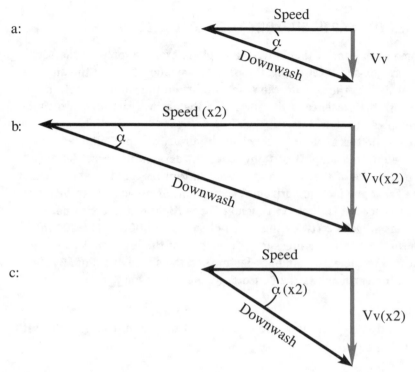

a:

Speed

α

Downwash

Vv

b:

Speed (x2)

α

Downwash

Vv(x2)

c:

Speed

α (x2)

Downwash

Vv(x2)

FIGURE 1.8   Effects of speed and angle of effective attack on downwash.

pellers would be a much less efficient means of propulsion. Only the component of thrust in the direction of motion of the airplane would be of use, and the rest of the thrust would represent wasted energy. Likewise, since the force of lift is up, one would expect the accelerated air that generates it to be down. Any other direction also would represent wasted energy.

The wing develops lift by transferring momentum to the air. Momentum is mass times velocity. In straight-and-level flight, the momentum is transferred toward the earth. This momentum eventually strikes the earth. If an airplane were to fly over a very large scale, the scale would weigh the airplane. The earth does not become lighter when an airplane takes off. This should not be confused with the (wrong) concept that the earth somehow supports the airplane. It does not. Lift on a wing is very much like shooting a bullet at a tree. The lift is like the recoil that the shooter feels, whether the bullet hits the tree or not. If the bullet hits the

DOES THE EARTH SUPPORT THE AIRPLANE?

Some people insist that since the airplane exerts a force on the earth, in straight-and-level flight, the earth is somehow holding the airplane up. This is definitely not the case. The lift on the wing has nothing to do with the presence of the surface of the earth. Examining two simple examples can show this.

The first example is to consider the thrust of a propeller, which is just a rotating wing. It certainly does not develop its thrust because of the presence of the surface of the earth. Neither could the presence of the earth provide the horizontal component of lift in a steep bank.

The second example to consider is the flight of the Concorde. It cruises at Mach 2 (twice the speed of sound) 55,000 ft (16,000 m) above the earth. The pressure information of the jet cannot be communicated to the earth and back faster than the speed of sound. By the time the earth knows the Concorde is there, it is long gone.

tree, the tree experiences the event, but that has nothing to do with the recoil of the gun.

The Adjustment of Lift

We have said that the lift of a wing is proportional to the amount of air diverted per time times the vertical velocity of that air. And we also have stated that in a rest frame where the air is initially at rest and the wing is moving, the air is moving almost straight down after the wing passes.

So what would happen if the speed of the wing were to double and the angle of attack were to remain the same? This is shown in Figure 1.8b. As you can see, the vertical velocity V_v has doubled. As we will soon see, the *amount of air diverted also has doubled*. As will be discussed shortly, the amount of air diverted is proportional to the speed of the airplane. Thus, in this case, *both the amount of air diverted and the vertical velocity of the air have doubled* with the doubling of the speed and keeping the angle of attack constant. Thus the lift of the wing has gone up by a factor of 4.

In Figure 1.8c, the wing has been kept at the original speed, and the relative angle of attack has been doubled. Again, the vertical velocity of the air has doubled, and since the amount of air diverted has not been affected, the lift of the wing has doubled. What these figures show is that the vertical velocity of the air is proportional to both the *speed* and the *angle of attack* of the wing. Increase either, and you increase the lift of the wing.

The pilot has controls for both airspeed and angle of attack. The airspeed is controlled by the power setting, along with the rate of climb or descent. The angle of attack is seen as a tilt of the entire airplane relative to the direction of flight and is controlled with the elevator, usually on the rear of the horizontal stabilizer, as shown in Figure A.1 (in Appendix A). The elevator works just like the wing in that it pushes air up or down to create a downward or upward lift, thus tilting the airplane.

Angle of Attack

Now let us look in more detail at the angle of attack of the wing. The *geometric angle of attack* is defined as the angle between the *mean chord* of the wing (a line drawn between the leading edge and the trailing edge of the wing) and the direction of the relative wind. This is what aeronautical engineers are referring to when they discuss angle of attack. For our discussion, we are going to use the *effective angle of attack*. The effective angle of attack is measured from the orientation where the wing has zero lift. The difference between the geometric angle of attack used by most people and the effective angle of attack used here should be emphasized to prevent potential confusion by the reader. Figure 1.9 shows the orientation of a cambered wing (see

> Harriet Quimby was the first U.S. woman to earn a pilot certificate, and in 1911 she was the first woman to pilot a plane across the English Channel.

also Figure A.4 and text) with zero *geometric* angle of attack and the same wing with zero *effective* angle of attack. A cambered wing at zero geometric angle of attack has lift because there is a net downward diversion of the air. By definition, the same wing at zero effective angle of attack has no lift and therefore no net diversion of the air. In the case of a symmetric wing, the geometric and effective angles of attack are, of course, the same.

For any wing, from that of a Boeing 787 to a wing in inverted flight, an orientation into the relative wind can be found where there is zero

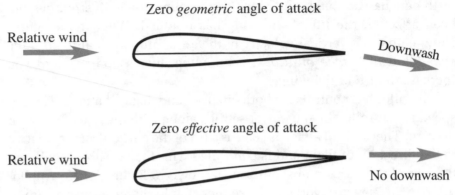

FIGURE 1.9 Definition of geometric and effective angles of attack.

lift. As stated earlier, this is zero effective angle of attack. Now, if one starts with the wing at zero degrees and rotates it both up and down while measuring the lift, the response will be similar to that shown on the graph in Figure 1.10. One can see that the lift is directly proportional to the effective angle of attack. The lift is positive (up) when the wing is tilted up and negative (down) when it is tilted down. When corrected for area and aspect ratio, a plot of the lift as a function of the effective angle of attack is essentially the same for all wings and all wings inverted. This is true until the wing approaches a stall. The stall begins at the point where the angle of attack becomes so great that the airflow begins to separate from the trailing edge of the wing. This angle is called the *critical angle of attack* and is marked in the figure. For two-dimensional (or infinite) wing simulations, lift as a function of effective angle of attack is identical for all airfoils.

Figure 1.10 also shows cross sections of the wings. A sharp, symmetric wing stalls earlier and more abruptly than a thick, asymmetric wing, but for smaller angles, the lift is the same for both. Turn the figure over, and you have the two wings' lift characteristics in inverted flight. Thus, as stated in the Introduction, any explanation of lift on a wing that depends on the shape of the wing is misleading at best. Such explanations also have trouble with explaining inverted flight, symmetric wings, and the adjustment of lift with load and speed. Again, it should be noted that the shape of the wing does affect the stall and drag characteristics of the wing.

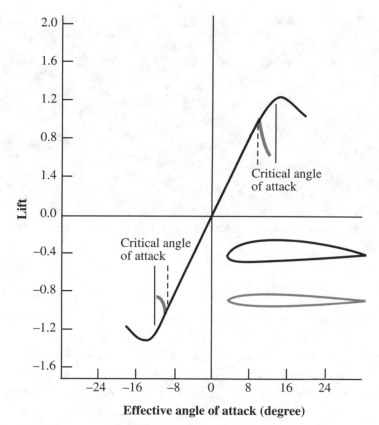

FIGURE 1.10 Lift as a function of angle of attack.

This is an extremely important result. It shows that the lift of a wing is proportional to the effective angle of attack. This is true for all wings: those of a modern jet, wings in inverted flight, a flat plate, or a paper airplane—or, for that matter, a bird's wing, as can be seen in the photo of a tern in Figure 1.11.

As can be seen in Figure 1.10, the relationship between lift and the angle of attack breaks down at the critical angle of attack. At this angle, the forces become so strong that the air begins to separate from the wing, and the wing loses lift while experiencing an increase in drag, a retarding force. At the critical angle, the wing is entering a stall. The subject of stalls will be covered in detail in Chapter 2.

FIGURE 1.11　Angle of attack of a tern in flight. (Photograph by Bernard Zee.)

The Wing as a "Virtual Scoop"

Newton's second law tells us that the lift of a wing is proportional to the amount of air diverted times the vertical velocity of that air. We have seen that the vertical velocity of the air is proportional to the speed of the wing and to the effective angle of attack of the wing. We have yet to discuss how the amount of air is regulated. For this, we would like to adopt the *visualization aid* of looking at the wing as a *virtual scoop* that intercepts a certain amount of air and diverts it to the angle of the downwash. This is not intended to imply that there is a real, physical scoop with clearly defined boundaries and uniform flow. However, this visualization aid does allow for a clear understanding of how the amount diverted air is affected by speed and density. The concept of the virtual scoop does have a real physical basis that will be discussed at the end of this section.

> The vertical velocity of the air is proportional to both the speed and the angle of attack of the wing.

For wings of typical airplanes, it is a good approximation to say that the amount of mass (quantity of air) diverted is proportional to the area of the wing. However, the amount of air diverted is not constant from tip to tip and falls off the farther you are from the wing surface. More air is diverted at the center of the wing and decreases at the wing tips. Thus we illustrate this with our conceptual virtual scoop having the shape shown in Figure 1.12.

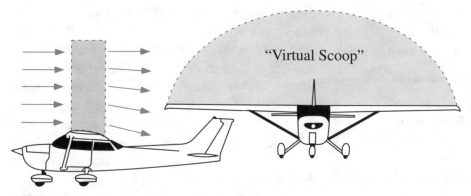

FIGURE 1.12   The virtual scoop as a visualization tool.

If one were to move such a virtual scoop through the air, how much air would it divert? Certainly, if the virtual scoop were to divert a certain amount of air at one speed, it would divert twice as much air at twice the speed. But what if the virtual scoop were taken to a higher altitude where the air has a lower density? If the air were half as dense, the virtual scoop would divert half as much mass for a given speed. Thus the amount of air intercepted by a wing is proportional to its area, the speed of the airplane, and the density of the air.

For the normal changes in angle of attack experienced in flight, we can make the statement that the amount of air diverted by the wing is independent of the angle of attack of the wing. The amount of air diverted also does not depend on the load on the wing, which is affected by the angle of attack.

Let us do a back-of-the-envelope calculation to see how much air might divert by a wing. Take, for example, a Cessna 172 that weighs about 2300 lb (1045 kg). Traveling at a speed of 140 mi/h (220 km/h) and assuming an effective angle of attack of 5 degrees, we get a vertical velocity for the air of about 11.5 mi/h (18 km/h) right at the wing. If we assume that the *average* vertical velocity of the air diverted is half this value, we calculate from Newton's second law that the amount of air diverted is on the order of 5 (English) tons per second. Thus a Cessna 172 at cruise is diverting about five times its own weight in air per second to produce lift. Think how much air a 600-ton Airbus A380 diverts.

Since we now know that a Cessna 172 diverts approximately 5 tons of air per second, let us get an idea of the size of our virtual scoop. If, for simplicity, we take the diverted air to be uniform and rectangular for the calculation, with a length equal to the wingspan of 36 ft (14.6 m), we get a height of about 18 ft (7.3 m). This is a lot of air. One should remember that the density of air at sea level is about 2 lb/yd³ (about 1 kg/m³). As implied by the shape of the virtual scoop in Figure 1.12, the lift is greatest at the root of the wing, tapering to the wingtip. Thus the air is diverted from considerably farther above the root of the wing than the 18 ft estimated here.

> On April 12, 1989, a British Airways Concorde landed in Sydney, Australia, with half its top rudder missing.

These estimates are not meant to be exact but rather to show the magnitude of the amount of air diverted by the wing. A great deal of air is diverted to produce lift. This is why flight is relatively efficient. As will be discussed later, the most efficient lift would be produced by diverting almost an infinite amount of air at almost zero vertical velocity.

Earlier it was mentioned that the pressure difference between the top and bottom surfaces of the wing is the mechanism by which the air

THE VIRTUAL SCOOP

The virtual scoop is introduced in this book for convenience. It is a visualization aid that helps one to understand how the amount of air diverted per second relates to the speed of the wing and to air density. The concept of the virtual scoop is loosely based on the theoretical aerodynamics law of Biot and Savart. This law gives the change in velocity around a wing by summing (or integrating) the influence of the various parts of the wing's surface. The influence of the wing on the air decreases with distance above the wing. Since the load on the wing is greatest at the root and decreases to near zero at the wing tip, the influence on the air, and thus the air diverted, is greatest at the root, decreasing to near zero at the wing tip. This variation in air diverted by the wing is reflected in the general shape of the virtual scoop.

The great strength in the virtual scoop as a visualization aid is that it, along with Newton's second law, yields the correct functional form for the variation of angle of attack and induced power with speed, load, and altitude. It also allows one to intuitively understand many complex phenomena, such as the power curve.

lifts the wing. This can give the impression that lift is a very local effect involving a small amount of air. One often sees pictures of airflow over a wing that show only a small amount of affected air, with "undisturbed air" a short distance above the wing. Our back-of-the-envelope calculation of the amount and extent of the air involved in the lift of a wing shows that this is not true. A great deal of air is involved in the production of lift. Focusing on surface pressures gives an incomplete picture of lift.

This large amount of diverted air causes the lower wing of a biplane to interfere with the lift of the upper wing. The air diverted by the lower wing reduces the air pressure on the bottom of the upper wing. This reduces the lift and efficiency of the upper wing. Thus many biplanes have the upper wing somewhat forward of the lower wing, or at least the root of the upper wing is moved forward to reduce this interference.

Putting It All Together

You now know that the lift of a wing is proportional to the amount of air diverted times the vertical velocity of that air. The amount of air diverted by the wing is proportional to the speed of the wing and the density of the air. The vertical velocity of the downwash is proportional to the angle of attack and the speed of the wing. This is shown in a tabular form in Table 1.1. With this knowledge, we are in a position to understand the adjustment of lift in flight.

As our first example, let us look at what happens if the load on a wing is increased by the airplane going into a 2g turn. In such a turn, the load on the wings has doubled. If we assume that the speed and altitude of the airplane are kept constant, the vertical velocity of the

TABLE 1.1 Relationship of Lift (Mass Diverted Times V_v) to Speed, Angle of Attack, and Air Density

	Speed	Angle of Attack	Air Density
Mass diverted	Proportional		Proportional
Vertical velocity V_v	Proportional	Proportional	

downwash must be doubled to compensate for the increased load. This is accomplished by doubling the angle of attack.

Now what happens when an airplane flying straight and level doubles its speed? If the pilot were to maintain the same angle of attack, both the amount of air diverted and the vertical velocity of the downwash would double. Thus the lift would go up by a factor of 4. Since the weight of the airplane has not changed, the increased lift would cause the airplane to increase altitude. Therefore, to maintain a constant lift, the angle of attack must be decreased to decrease the vertical velocity. If the speed of an airplane were increased by 10 percent, the amount of air diverted by the virtual scoop and the downwash would both increase by about 10 percent. Thus the effective angle of attack would have to be decreased to 82.5 percent ($1.1 \times 1.1 \times 0.825 = 1$). The lift of the wing then would remain constant.

As our last example, let us consider the case of an airplane going to a higher altitude. The density of the air decreases, and so, for the same speed, the amount of air diverted has decreased. To maintain a constant lift, the angle of attack has to be increased to compensate for this reduction in the diverted air. If the density of the air is reduced by 10 percent, the vertical velocity of the downwash has to be increased by 11 percent to compensate. This is accomplished by increasing the effective angle of attack by that amount.

> The German aeronautical pioneer Otto Lilienthal died after his glider stalled and crashed. As he lay dying, he was quoted as saying, "Sacrifices must be made."

We now understand how the airplane adjusts the lift for varying load, speed, and altitude. Now we must understand the roles of power and drag to round out our understanding of flight. We will start with a look at power.

Power

One of the most important concepts for understanding flight is that of the power requirements. In aeronautics textbooks, the discussion of *drag*, which is a force against the motion of the airplane, would come first, and power would be given little consideration. This may be appropriate for the design of an airplane, but it is less useful to a pilot trying to understand operating an aircraft.

Power is the rate at which work is done. Or, in simpler terms, power is the rate fuel is consumed. The power associated with flight also relates to the demand placed on the engine and the limitations on airplane performance. We will consider two types of power requirements. The first is *induced power*, which is the power associated with the production of lift. It is equal to the rate at which energy is transferred to the air to produce lift. Thus, when you see the word *induced* with respect to flight, think of lift. The second power requirement we need to consider is *parasite power*. This is the power associated with the impact of the air against the moving airplane. The *total power* is simply the sum of the induced and parasite powers.

> The B-17 in World War II shot down nearly half the total enemy airplanes downed by U.S. aircraft, and it was a bomber, not a fighter.

Induced Power P_i

Let us first look at the induced power requirement of flight. The wing develops lift by accelerating air down. Before the wing came by, the air was standing still. After the wing passes, the air has a downward velocity, and thus it has been given kinetic energy. *Kinetic energy* is energy associated with motion. If one fires a bullet with a mass m and a velocity v, the energy given to the bullet is simply $\frac{1}{2}mv^2$. Since the induced power is the rate at which energy is transferred to the air, it is proportional to the *amount of diverted air* times the *vertical velocity squared* of that air. (Remember that in the rest frame of the observer on the ground, the direction of the downwash is essentially down.) However, since the lift of a wing is proportional to the amount of air diverted times the vertical velocity of that air, we can make a simplification. *The induced power associated with flight is proportional to the lift of the wing times the vertical velocity of the air.* To put this in an equation form to make it easier to understand, we can write

$$P_i \sim \text{mass per time} \times V_v^2$$

$$\alpha\, L \times V_v$$

where the symbol ~ means "proportional to" and L is the lift on the wing.

Now let us look at the dependence of induced power on the speed of the airplane. We know that if the speed of an airplane were to double, the amount of air diverted also would double. Therefore, the angle of attack must be adjusted to give half the vertical velocity to the air, maintaining a constant lift. The lift is constant, and the vertical velocity of the downwash has been halved. Thus the induced power has been halved. From this we can see that the *induced power varies as 1/speed for a constant load.* The induced power is shown as a function of speed by the dotted line in Figure 1.13. This shows that the more slowly the airplane flies, the greater is the power requirement to maintain lift. As the airplane slows in flight, more and more power must be added, and the angle of attack must be increased. Finally, the airplane is flying at full power with the nose high in the air. What is happening is that as the airplane's speed is reduced, more and more energy must be given to less and less air to provide the necessary lift.

Parasite Power P_p

Parasite power is associated with the energy lost by an airplane to collisions with the air. It is proportional to the average energy that the airplane transfers to an air molecule on colliding times the rate of collisions. As with the energy given to the bullet mentioned earlier, the energy lost to the air molecules is proportional to the airplane's speed squared. The rate of collisions is simply proportional to the speed of the airplane. The faster the airplane goes, the higher is the rate of collisions. Thus we have a speed-squared term owing to the energy given to

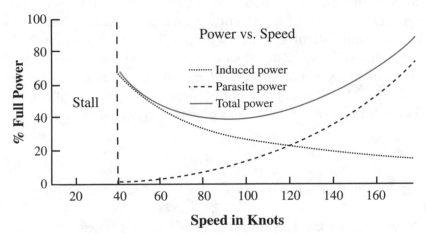

FIGURE 1.13   The power required for flight as a function of speed.

each molecule through collisions and a single speed term owing to the collision rate. This yields the result that the *parasite power varies as the speed cubed*. This means that if the speed of the airplane were to increase by a factor of 2, the parasite power would increase by a factor of 8! The parasite power as a function of speed is also graphed in Figure 1.13 by the second dashed line.

Can a person in free fall reach a supersonic speed? At 10,000 ft, a person would have to be traveling at 750 mi/h (1230 km/h). A person in a dive would have difficulty reaching even 500 mi/h (820 km/h).

The fact that the parasite power goes as the airplane's speed cubed has an important consequence for the performance of an airplane at cruise speed. Here, the power requirements are dominated by the parasite power. In order for an airplane to double its cruise speed, it would have to increase the size of its engine by almost eight times! Thus, when an airplane owner upgrades to a larger engine, there is an improvement in the rate of climb and turn of the airplane but only a modest increase in cruise speed. To substantially increase the speed of the airplane, the parasite power requirement must be decreased. Such design features as retractable landing gear, smaller fuselage cross sections, and an improved wing design accomplish this.

Power requirements are associated with the rate of fuel consumption or the loss of altitude. At cruise speeds, power is dominated by parasite power, which goes as the speed cubed. Therefore, the rate of fuel consumption also goes as the speed cubed. However, since the distance travel per unit time is also increasing, the fuel consumption in units of distance per unit volume (say, miles per gallon) goes as the speed squared. Thus the work that the engine has to do goes as speed cubed, and the fuel efficiency decreases as the speed squared. This is why a car that seems to have plenty of power at fast highway speeds still has a fairly low top-end speed and slowing down on the highway greatly increases the car's efficiency.

The Power Curve

As stated earlier, total power is the sum of the induced and parasite powers. The solid colored line in Figure 1.13 shows the total power as a function of speed. At low speed, the power requirements of an airplane are dominated by the induced power, which goes as 1/speed. At cruise speeds, the performance is limited by the parasite power, which goes as the speed cubed. This graph of total power as a function of speed is known as the *power curve*. Flying at slow speeds, where the total power requirement increases with decreasing speed, is what pilots refer to as flying the *backside of the power curve* or, in pilot training, as *slow flight*.

One might ask how an increase in altitude would affect the power curve. This is illustrated in Figure 1.14, which shows the power curves for altitudes of 3000 and 12,000 ft (about 900 m and 3600 m). With an increase in altitude, there is a decrease in air density. Thus the wing diverts less air, and the angle of attack must be increased to maintain lift. Since the induced power is proportional to V_v, which must be increased with altitude at a fixed speed, induced power also increases with altitude. Going from 3000 to 12,000 feet represents approximately a 30 percent reduction in air density, and thus there is a compensating 40 percent increase ($1/0.7 = 1.4$) in V_v and induced power. An airplane flying on the backside of the power curve would require more power and fly with a greater angle of attack when going to a higher altitude. Note that since the critical angle of attack does not change with air density, the stall speed also has increased by 40 percent at 12,000 feet. That is, the airplane will stall at a higher speed.

The situation is the opposite for parasite power. A reduction in air density translates to a reduction in the number of collisions with the

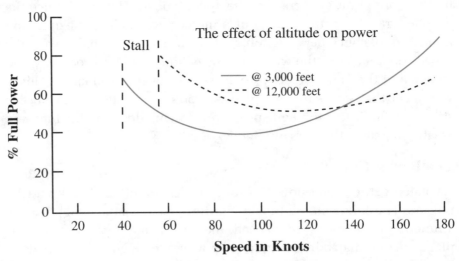

FIGURE 1.14   Total power required for flight at two altitudes.

air, and thus there is a reduction in parasite power. Therefore, the reduction by 30 percent in air density by climbing from 3000 to 12,000 feet gives a 30 percent decrease in parasite power losses. An airplane at cruise speed where parasite power dominates finds it more economical to fly at a higher altitude.

Usually flying at a higher altitude does not translate into flying at a higher speed because nonturbocharged engines experience a reduction in power that is similar to the reduction in atmospheric density. That is, if the atmospheric density is 65 percent that of sea level, the maximum power of the engine is also approximately 65 percent of its sea-level performance. This is easy to understand because a reduction in air density is a reduction in oxygen and thus a reduction in the amount of fuel that can be burned.

> Before 1900, Langley's law, found experimentally by Samuel Langley, said that total power required for flight decreases with increasing speed. It has been shown that all of Langley's experiments were performed on the backside of the power curve.

Effect of Load on Induced Power

Now let us examine the effect of load on induced power, that is, the power requirement to produce the needed lift. First, remember that the induced power associated with flight is proportional to the *lift* of the wings times the *vertical velocity* of the downwash. Now, if we were to double the load, maintaining the same speed, we would have to double the vertical velocity of the air to provide the necessary lift because the rate that air is diverted has not changed. Both the load and the vertical velocity of the air have been doubled, and the induced power has gone up by a factor of 4. Thus, at a constant speed, the *induced power increases as the load squared*. It is easy to see why the weight of an airplane and its cargo is so important. Figure 1.15 shows the data for the relative fuel consumption of a heavy commercial jet as a function of weight. These measurements were made at a fixed speed. From the data, one can estimate that at a gross weight of 500,000 lb (227,000 kg) and a speed of Mach 0.6 (60 percent of the speed of sound), about 40 percent of the power consumption is induced power and 60 percent is parasite power. In reality, the airplane would cruise at a speed of around Mach 0.8, where the induced power would be lower and the parasite power consumption would be higher. Unfortunately, at this more realistic speed, the details of power consumption become more

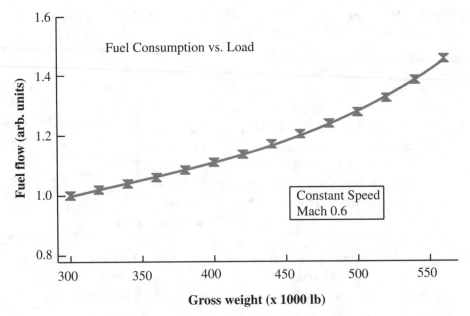

FIGURE 1.15 Fuel consumption as a function of weight for a large jet at a constant speed.

complicated, and it is more difficult to separate the parasite and induced powers from the available data.

Drag

So far we have discussed power at length, with only brief references to the topic of drag. With an understanding of power, we are in a position to understand drag, which is part of the pilot's culture and vocabulary. Drag is a force that resists the motion of the airplane. Clearly, a low-drag airplane will fly faster than a high-drag airplane. It also will require less power to fly the same speed. So what is the relationship among power, drag, and speed?

Power is the rate at which work is done. It also can be looked at as the rate fuel is being consumed. In mathematical terms, it is also force times velocity. *Drag is a force and is simply equal to power divided by speed*. We already know the dependence of induced and parasite powers on speed. By dividing power by speed, we have the dependence of drag on speed. Since induced power varies as 1/speed, *induced drag*

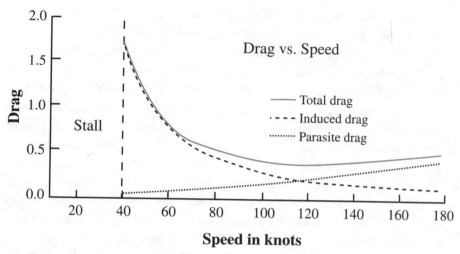

FIGURE 1.16   Drag as a function of speed.

varies as 1/speed squared. Parasite power varies as speed cubed, so *parasite drag varies as speed squared*. Figure 1.16 shows the dependence of induced, parasite, and total drag on the speed of the airplane. The dependence of power and drag on speed is summarized in Table 1.2.

In the preceding section we saw that the induced power increases as load squared. Since drag is just power divided by speed, induced drag also increases as load squared. Anything understood about power can be easily converted to a similar understanding of drag simply by dividing by speed.

We have said that drag is part of a pilot's culture and vocabulary. This is true. But most of the time when the term is used, the person really means *power*. Let us look at an example to illustrate. Take the case of a pilot flying a small plane with retractable landing gear. If full power is applied in straight-and-level flight with the gear up, the air-

TABLE 1.2   Dependence of Power and Drag on Speed

	Power	Drag
Induced	$1/\text{speed}$	$1/\text{speed}^2$
Parasite	Speed^3	Speed^2

plane accelerates to some speed and goes no faster. The pilot might well say that the airplane's speed is limited by the drag.

Let us imagine that the airplane has two meters, one that measures total drag and another that measures the total power for flight. We then will record both values for the airplane at its top speed. The pilot lowers the landing gear and flaps, leaving the engine at full power. There is now a substantial increase in the drag and power required. This, of course, slows the airplane down. We would find that the airplane slowed down to the previous total power requirement, and now, the total drag is considerably higher than before. The pilot would have to reduce power to bring the total drag down to its previous value. Its top speed was not determined by the total drag but by the total power. Thus, when pilots say drag, they often mean power.

The utility of the concept of power over drag for a pilot is fairly easy to understand. Power requirements relate simply to the demands on the engine. Drag is a force that must be related to the airplane's speed in order to understand the power requirement to overcome it. A drag of a certain value at one speed is only half the power drain of the same drag at twice the speed. In the end, the power available from the engine is what counts.

The Wing's Efficiency for Lift

We have seen that the induced power requirement of a wing varies as 1/speed for a fixed load and as load squared for a fixed speed. But one may wonder how the design of the wing affects the induced power requirements. In other words, what is the wing's efficiency for lift?

Efficiency for lift has to do with the amount of induced power it takes to produce a certain lift. The lower the induced power needed, the greater is the efficiency. The most obvious way to improve the efficiency of a wing is to increase the amount of air diverted by the wing. If more air is diverted, the vertical velocity of the air is reduced for the same lift and so is the induced power. This can be accomplished by increasing the size of the wing. One can increase the wing area by either increasing the span or increasing the chord, which changes the aspect ratio of the wing.

> The Wright brothers' first flight at Kitty Hawk, North Carolina, could have been performed within the 150-ft (45-m) economy section of a Boeing 747-400.

For a simple rectangular wing, the *aspect ratio* is the span (tip-to-tip distance) divided by the chord (leading-edge-to-trailing-edge distance). Generically, it is the span divided by the mean chord or, more accurately, the wing area divided by the span squared.

If two wings have the same area, the longer one will be the most efficient. This is so because as the chord of a wing becomes large, it becomes less effective at diverting air. Gliders operate at speeds where induced power dominates. Thus high-performance gliders have very high aspect ratio wings. Figure 1.17 shows a glider with a 60:1 glide ratio. This means that in still air, the glider travels 60 ft horizontally for every foot it descends to provide power. It is interesting to note that sea birds, which must fly long distances without landing, also have high-aspect-ratio wings for optimal efficiency. Birds of prey have low-aspect-ratio wings for strength during maneuvering.

In consideration of the total efficiency of a wing, the parasite power also must be considered. The parasite power of a wing is proportional to its surface area. Thus, for cruise speeds where parasite power dom-

FIGURE 1.17   A high-performance glider. (Photograph courtesy of Motorbuch Verlag, Stuttgart, Germany, from Cross-Country Soaring.)

inates, there is a limit to how much the area of the wing can be increased to reduce the induced power. There are additional problems with increasing the wing's area, particularly by increasing its span. The first is that large wings are heavy and increase the weight of the airplane. The second is that long wings are not as structurally strong as shorter wings.

Most fast airplanes have shorter wings. An exception is the U-2 spy plane, which flew at 460 mi/h (740 km/h) above 55,000 ft (16,700 m). The U-2 (see Figure 2.4) had long wings because of the extreme altitudes at which it operated. Because of the low density of the air, the induced power was significant at its cruise speed.

> The first around-the-world airplane flight occurred from April 6 to September 28, 1924, and started and ended in Seattle.

Because the rotors on a helicopter are quite small for the weight of the aircraft, they must accelerate a relatively small amount of air to a high velocity (high kinetic energy) to produce the needed lift. This is less efficient. A similar argument can be made to understand why an airplane cannot "hang" on its propeller. Although the engine is producing sufficient power to lift the airplane with the wings, the propeller accelerates too little air at too high a velocity to produce the necessary lift on its own. One airplane that can hang on its propellers is the Bell-Boeing V-22 Osprey tilt-rotor aircraft shown in Figure 1.18. In the figure, the aircraft is transitioning from vertical flight to forward flight. The extreme propellers, or *proprotors*, divert a great deal of air, allowing the engines to pro-

FIGURE 1.18 The V-22 Osprey. (Photograph courtesy of the U.S. Air Force.)

LIFT REQUIRES POWER

In classic two-dimensional aerodynamics, one learns that a wing requires no work to produce lift and that there is no net downwash behind the wing. This is illustrated through theory and calculations that are done with two-dimensional airfoils. These two-dimensional airfoils are in fact infinite wings. This is so because a two-dimensional, or infinite, wing is much easier to calculate than one of finite span. The efficiency of a wing increases with the span of the wing because the amount of air diverted increases with area. Thus an infinite wing diverts an infinite amount of air at near-zero velocity to produce lift and thus is infinitely efficient. The net vertical velocity of the downwash is essentially zero. Therefore, the infinite wing requires no power to produce lift. Of course, this is not the situation for a real three-dimensional wing.

Some people argue that there is no work done in producing lift. Work is force times distance. The argument goes that since the wing in straight-and-level flight does not change altitude, no work is done against the downward force of gravity. However, if you push a heavy box along a flat floor, you must do work against friction. Induced drag is the horizontal force produced as a consequence of lift, which, like friction in the box example, is opposite the direction of motion. So the work done in producing lift is the induced drag times the distance the wing travels.

duce enough thrust to lift the craft vertically off the ground before they are rotated forward for horizontal flight.

Wing Vortices

The downwash behind a wing is sometimes called the *downwash sheet*. This downwash sheet has a curl in it, producing the *wing vortex*. Near the tip of the wing, the wing vortex curls very tightly, creating the wingtip vortex. The wingtip vortex itself contains only a small portion of the total energy of the wing vortex. Because of the tight curl, it behaves like a small tornado with very low pressure in its core. Low pressure is tied to lower temperature, and so, under the right conditions, moisture condenses, making the wing tip vortex visible.

Eventually, the entire wing vortex curls into a single trailing vortex on each side. Since the strength of the trailing vortices is related to lift, a heavy aircraft produces a stronger vortex than a light aircraft. When trailing a large aircraft for landing, trailing aircraft must leave enough separation to allow the trailing vortex to sink and dissipate. Otherwise, the trailing aircraft could get flipped by the preceding trailing vortex. The wing vortex sinks because in the rest frame of the ground, any piece of it is traveling almost straight down, as discussed earlier.

To understand why the wing vortex curls, we must first consider the lift distribution of the wing. In our discussion of the virtual scoop, we illustrated in Figure 1.12 that the amount of air diverted by a wing is maximum near the root and decreases to zero at the wingtip. The height of the virtual scoop at any point along the wing represents the load and the momentum transferred at that point. The load on a wing is nicely illustrated in Figure 1.19, which shows the condensation on top of the wing of a fighter aircraft during a high-*g* maneuver. Just as in the wingtip vortex, the lowered pressure above the wing reduces the air temperature, causing condensation in the form of a kind of fog. This fog displays the variation in load along the wing. One can see that the load is greatest near the root of the wing, and it tapers to zero at the wingtip. The large fuselage causes the condensation to decrease along the centerline of the fighter.

In the section on air bending over a wing, we saw that faster-moving air bends toward slower-moving air. The same thing happens with

FIGURE 1.19   Illustration of the lift distribution on the wings of an F-14A fighter jet. (Photograph courtesy of NASA.)

the downwash sheet. The velocity is highest at the root and slows down as one progresses to the wingtip. The result is that after the wing has passed, the downwash at the center of the sheet bends toward the wingtip. Eventually, as mentioned earlier, the entire downwash sheet rolls up into two counterrotating vortices, one on each side of the fuselage. The details of the wing vortex can be seen clearly in Figure 1.7 of a jet flying over fog. The center of the trough formed by the downwash sheet has smooth sides terminating in the tight curl of the wingtip vortex.

The tightness of the curl is related to the rate of change of downwash velocity (i.e., lift) as one moves along the wing. The wingtip is usually the place on the wing with the greatest change in lift and thus produces the most visible vortex. This is not always the case. Figure 1.20 shows a landing airplane producing *flap vortices*. In this example, the change in lift is greatest at the outer edge of the wing's flap.

> The first flight attendants (i.e., stewardesses) were required to be registered nurses.

Circulation

Circulation is a measure of the rotation of the air around a wing when seen from the rest frame, where the air is initially standing still and the

FIGURE 1.20   Wing flap vortices. (Photograph by Jan-Olov Newborg.)

wing is moving. Circulation has been employed mistakenly by some as a "driving" mechanism to accelerate the air over the top of the wing and thus account for the reduction in pressure causing lift. Let us go back to the rest frame of an observer on a mountaintop who is able to take a picture of the directions of air movement around a wing as it passes. What would the picture look like? It would look something like Figure 1.21. Remember when studying this figure that it is a snapshot of a moving wing and that the arrows represent the velocities in the air at one moment in time. The air is not moving around the wing but is shifting in a circular pattern. The arrow marked "1" will become the arrow marked "2" in a moment, and so on. If one adds the speed and direction of the relative wind (as seen by the wing) to each of the arrows in the figure, the familiar streamlines with upwash and downwash are produced.

So why is this happening? First, we have to bear in mind that air is considered an incompressible fluid for these discussions. This means that it cannot change its volume and that there is a resistance to the formation of voids. Now the air has been accelerated over the top of the wing by the reduction in pressure. This draws air from in front of the wing and expels it back and down behind the wing. This air must be compensated for, so the air shifts around the wing to fill in. This is

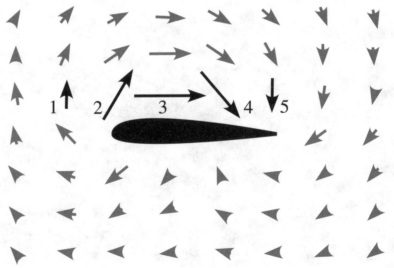

FIGURE 1.21   Airflow around a wing as seen by an observer watching on the ground.

similar to the circulation of the water around a canoe paddle. This circulation around the wing is no more the driving force for the lift on the wing than is the circulation in the water that drives the paddle. It is true that if one is able to determine the circulation around a wing, the lift of the wing can be calculated. Lift and circulation are related to each other.

The North American P-51 Mustang prototype was designed and built in 102 days.

The most obvious result of the circulation around the wing is that the air approaches from below the wing. This is the cause of upwash. When an airplane is traveling at Mach 1 or faster, information cannot communicate forward, and the wing has no upwash. However, at the speed common for small airplanes, upwash is quite pronounced. If Figure 1.21 were expanded so that more of the air could be seen, one would see that the air is going almost straight down far behind the wing and that the effect of the wing extends far above the wing.

Something to notice in the simulations of circulation is that there is very little action below the wing. Most of the action is occurring above

FIGURE 1.22   **Armament goes on the bottom of the wing of an F-16. (Photograph courtesy of the U.S. Air Force.)**

the wing. This is why the bottom side of military aircraft can be so cluttered with munitions and fuel tanks, as in the F-16 fighter shown in Figure 1.22. These obstructions cause an increase in parasite power losses but do little to affect the efficiency of the wing. The top of the wing is a different story. Obstructions above the wing interfere with lift. This interference with lift explains why struts are common on the bottoms of wings but are historically rare on the tops of wings.

Ground Effect

The concept of *ground effect* is well known to pilots. This effect is the increase in efficiency of a wing as it comes to within about a wing's length of the ground. The effect increases with reduction in the distance to the ground. A low-wing airplane will experience a reduction in the induced drag of as much as 50 percent just before touchdown. This reduction in drag just above a surface is used by large birds, which often can be seen flying just above the surface of a body of water. Pilots taking off from deep-grass or soft runways also use ground effect. The pilot is able to lift the airplane off the soft surface at a speed too slow to maintain flight out of ground effect. This reduces the resistance on the wheels and allows the airplane to accelerate to a higher speed before climbing out of ground effect.

What is the cause of this reduction in drag? There are two contributions that can be credited with the reduction in drag. The ground influences the flow field around the wing, which, for a given angle of

FLIGHT OF INSECTS

Statements have been made that classic aerodynamic theory proves that insects cannot fly. In all cases that we are familiar with, the flight of insects is expressed in terms of circulation. Circulation is a model developed for large aircraft and does not apply well to small insects. Insects obey the same laws of physics as airplanes. They produce lift by blowing air down. When you have a chance, observe a bumblebee pollinating flowers. You will see that when it flies over a leaf on the plant, the leaf is depressed just as if it had landed on it. It clearly is producing lift the same way an airplane does.

attack, increases the lift. At the same time, though, there is a reduction in downwash. It can be surmised that this additional lift must come from an increase in pressure between the wing and the ground. In addition, since lift is increased for a given angle of attack, the angle of attack can be reduced for the same lift, resulting in less downwash and less induced drag.

Ground effect introduces a fundamental change from the discussion of flight at altitude. When no ground is present, the relationship among lift, drag, and downwash is straightforward. Near the ground, however, there is an action-reaction among the wing, the air, and the ground. At altitude, the ground is so distant that this effect does not exist. Near the ground, this interaction helps to produce lift and reduce downwash owing to an increase in pressure below the wing. The details of ground effect are extremely complex. Most aerospace texts devote a paragraph or two and do not attempt to describe it in depth. The truth is that so much is changing in ground effect that it is difficult to describe by pointing to a single change in the airflow or a term in an equation. There is no simple way to describe how the airflow adjusts to satisfy the change in conditions.

> The principle of equal transit times is not true for a wing with lift.

Lift on a Sail

It is sometimes asked if the physics of "lift" on a sail are different from those of lift on a wing. Of course, a sailboat sailing downwind is being pushed by the wind, and thus the physics are clearly different. However, a sailboat tacking at an angle of, say, 45 degrees to the wind is obeying the same physical principles as a wing.

Take a look at the shape of the sail in Figure 1.23, and then look at the wing of the Curtis aircraft in Figure I.2. You will see that they are essentially the same in shape as they are in function. The direction of the flow of air is changed, causing a force perpendicular to the initial direction of the wind. As discussed below, this initial direction of the wind is the apparent direction because the motion of the boat must be added to the motion of the air. As shown by the wind-velocity diagram in the lower right-hand corner of Figure 1.23, the air is bent by the sail, yielding a change in velocity that is perpendicular to the initial direc-

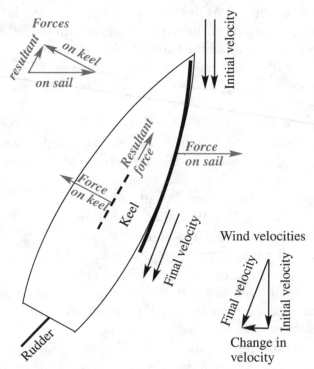

FIGURE 1.23 Forces and wind velocities on a sailboat.

tion of the air. This produces a force on the sail opposite this change in direction. In this case, much of the force is perpendicular to the direction of travel of the boat.

This is where the keel comes into play. The keel is a wing flying through the water. The angle of attack is controlled by the rudder, just as the elevators control the angle of attack of the wing. The keel is much smaller than the sail, and the angle of attack is small. The force produced is sufficient because the keel is "flying" in a fluid almost 1000 times more dense than air. The force on the keel is perpendicular to the direction of the boat in the water and produces a force that subtracts from the force on the sail in such a way that the resulting force is in the direction of motion of the boat. This is illustrated by the force diagram in the upper left-hand corner of the figure.

Like the wing of the Curtis airplane, some of the lift is produced by the air diverted by the concaved surface of the sail. As with a wing,

It is worth saying a few words about the effect of motion on the apparent wind velocity and direction. Take the example of a boat tacking at 45 degrees to a 10 km/h wind. In this example, the boat is traveling at 12 km/h across the water, as illustrated in Figure 1.24. The apparent wind seen by the boat is approximately 20 km/h. Also, the angle off the bow is only 18 degrees, not 45 degrees. As we will see later, understanding this change in direction and speed of the wind seen by a moving object is key to understanding autorotation (controlled descent of a helicopter without power) and wind generators.

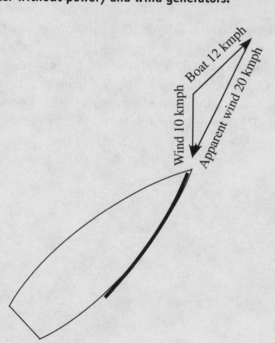

FIGURE 1.24   Velocity and direction of the apparent wind.

though, most of the lift is produced by the air that is accelerated by the lowering of the pressure by the convex surface.

Wrapping It Up

The key thing to remember about lift is that it is a reaction force caused by the diversion of air down. The physics of diverting the air down are expressed by the lowering of the pressure on the top of the wing producing the lifting force. The lift of a wing is proportional to the *amount* of air diverted times the *vertical velocity* of that air. For a given wing, the amount of air diverted is proportional to the speed of the wing and the *density* of the air. The vertical velocity of the downwash is proportional to the *angle of attack* and the *speed* of the wing. The induced power is proportional to the load times the vertical velocity of the downwash and varies as 1/speed. The parasite power is proportional to air density and varies as the speed cubed. With these basic concepts, the phenomenon of flight can be easily understood.

> Artist and inventor Leonardo da Vinci wrote 35,000 words and made 500 sketches concerning flight.

In this chapter we have downplayed the importance of the shape of the wing. It is obvious to anyone who has been around airplanes that wings are in fact very complicated structures with a great deal of engineering involved. Chapter 2 considers the wing in detail.

Wings

In the Introduction we saw that the shape of the wing was not the determining factor in lift. In Chapter 1 we presented a physical description of flight. This chapter will show how the concepts used to understand flight can help to understand the design of a wing. There are many factors that go into the design of a wing. Should the wing be swept back and tapered? What airfoil should be used? What high-lift devices should be added to improve takeoff and landing performance? These are just some of the questions that must be answered when designing a wing.

Besides aerodynamic considerations, the wing designer must consider other tradeoffs such as structural weight and cost. Some aerodynamically sound principles have fallen prey to the realities of construction costs, structural weight, or maintainability. Understanding these tradeoffs is more a function of experience than of formal training. Hopefully, reading this chapter will give you an appreciation for the decisions that must be made.

Airfoil Selection

Before a wing can be designed, a wing section, or airfoil, must be selected. As stated in Appendix A, the airfoil is a slice of a wing as viewed in cross section. In Chapter 1 it was emphasized that lift is primarily a function of the angle of attack, with little dependence on airfoil shape. So why will not almost any wing section do? Here, we will discuss some of the specific airfoil design characteristics that are used and how they affect performance. Characteristics that must be considered in selecting a wing include lift at the cruise angle of attack, drag, stall characteristics, laminar flow, and room for fuel and internal structures.

Wing Incidence

The lowest drag for the fuselage will be achieved when the fuselage is aligned with the relative wind. A wing is attached at an incident angle, or fixed effective angle of attack, so that the fuselage is at a zero angle to the wind while the wing is at some positive angle of attack. The *incident angle* is defined as the angle between the wing's chord and the longitudinal axis of the aircraft. The angle of incidence is usually just a few degrees in general aviation.

> The wristwatch became popular when Louis-Francois Cartier made one for aviator Alberto Santos Dumont to make it easier to tell time during the rigors of flying an air machine.

As will be seen below, the optimal speed of a large jet is designed into the wing. Therefore, the speed is fixed. The incident angle is set for a specific speed-load-altitude combination. Unfortunately, this is seldom met, so large jets almost always have some small misalignment to the relative wind. In an ideal world, large aircraft would be able to adjust the incident angle in flight, but that would not be practical because of weight and cost considerations.

Wing Thickness

Wing thickness is another design consideration. A thick wing can result in a large wake, resulting in high parasite drag, even at zero angle of attack. The airflow around a thick wing also may separate, causing *form drag*, a type of parasitic drag.

A thicker wing has a structural advantage because the wing's structure can be contained inside the wing itself. The thin wings used by early airplanes required external bracing and thus produced higher parasite drag. Before aerodynamic drag was understood toward the end of World War I, emphasis was placed on thin wings, with many wires and struts to give them strength. Biplanes were used most frequently because they made for a nice boxlike structure. There is a natural competition between the aerodynamicist who wishes to have a thin wing and the structural engineer who wants the wing to look internally like a nice fat box.

Leading Edge and Camber

The *camber* of a wing (see Figure A.4 in Appendix A) is a measure of the asymmetry between the curve of the upper and lower surfaces of the wing. A positively cambered wing has a kind of bulging top surface with respect to the bottom surface. Camber and the shape of the leading edge are often related and affect when and how a wing stalls. In general, the greater the camber, the greater is the lift before stall begins, and thus the greater is the load that the wing will support. This is illustrated in Figure 1.10.

The radius of curvature of the leading edge also affects the abruptness of stall entry. A wing with a sharp leading edge will go abruptly into a stall. A blunt leading edge will have a much less abrupt stall entry. The *separation point*, the point where the airflow begins to separate from the wing, smoothly moves from the trailing edge forward as the angle of attack is increased. As the separation point progresses forward, the stall becomes deeper. Figure 2.1 shows the early stall development of a wing with a well-rounded leading edge as well as the stall of a wing with a sharp leading edge. The wing with the sharp leading edge goes directly into a full stall. Another way of looking at stall entry is shown in Figure 1.10, which shows lift as a function of effective angle of attack of a well-cambered wing and a thin wing. It is clear that the sharper wing goes from maximum lift to full stall with a very small change in angle of attack.

Airplanes that are designed to operate at lower speeds or are used as trainers have fairly round leading edges. Fast jet fighters have sharper leading edges. A difficult problem for pilots transitioning from lower-speed trainers to high-speed fighters is appreciating the more abrupt stall. Many new pilots have been surprised when their high-speed airplane stalled without warning. A classic use of a sharp-leading-edge wing section is on the F-104 Star Fighter, shown in Figure 2.2. The leading edge is so sharp that it has been known to cut the hands of mechanics who inadvertently rub against it.

> The first "air raid" occurred in Libya as Italian Lieutenant Giulio Gavotti dropped four 4.4-lb grenades on Turkish positions.

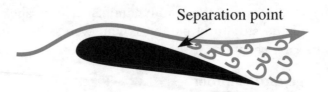

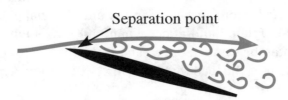

FIGURE 2.1 Flow separating from two different airfoils.

Short-takeoff-and-landing (STOL) aircraft have very fat, round leading edges. This will hurt the airplane's top cruise speed, but top cruise speed is usually not the primary goal of STOL aircraft. Figure 2.3 shows the wing of a typical STOL aircraft. It is much fatter than that of a typical general aviation airplane and has a well-rounded leading edge.

Airplanes that fly close to the speed of sound, where most commercial airplanes fly, experience supersonic flow over part of the upper surface of the wing. To reduce the drag associated with tran-

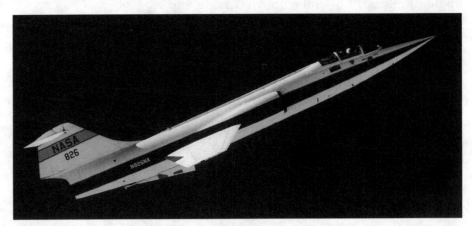

FIGURE 2.2 The F-104 Star Fighter with a thin, sharp wing. (Photograph courtesy of NASA.)

FIGURE 2.3 A STOL airplane with a fat leading-edge wing. (Photograph courtesy of Czech Aircraft.)

sonic and supersonic flow, which will be discussed in Chapter 5, the top surface is generally fairly flat, with most of the curvature on the lower surface of the airfoil. This is illustrated in Figure I.4 with the Whitcomb Supercritical Airfoil [NASA/Langley SC(2)-0714]. Selecting the right airfoil clearly depends on the mission for which the airplane is designed.

Wing Planforms

Along with selecting the right airfoil, or wing section, the designer also must select the right *planform*. The planform is the contour of the wing as viewed from above. How big should the wing be? Should it have a high or a low aspect ratio? Should the wing be swept? Should it be tapered? In this section we shall examine how to choose the basic planform of the wing.

Wing Loading

The first design parameter to determine is wing loading. This is the ratio of wing area to weight of the airplane, measured in pounds per square foot (lbs/ft²) or newtons per square meter (N/m²). Many basic performance parameters are determined as a function of wing loading, which will be discussed in Chapter 6. Wing loading will determine cruise performance, takeoff and landing distances, and power requirements. Typical wing loadings for various aircraft are shown in Table 2.1. Note that

TABLE 2.1 Typical Wing Loading

| | Wing Loading | |
Type	lb/ft^2	N/m^2
Sailplane	6	30
General aviation (single engine)	17	85
General aviation (twin engine)	26	130
Jet fighter	70	350
Jet transport	120	600

light trainers have low wing loading, whereas commercial transports and military aircraft have a wing loading as much as 10 times greater.

There are many tradeoffs to be considered with wing loading. One consideration is that the higher the wing loading, the higher is the stall speed. This is why trainers have lower wing loading. However, low wing loading also limits top cruising speed. A higher wing loading is desired for faster airplanes. Also, airplanes with higher wing loading are less susceptible to clear-air turbulence. The inertia of the airplane against the gusts of turbulence makes the airplane harder to blow around. Details of these tradeoffs will also be discussed in Chapter 6.

Aspect Ratio

The aspect ratio of the wing is the span (measured from tip to tip) divided by the mean chord. Chapter 1 discussed the advantage of high aspect ratio (long, narrow wings) for low-speed aircraft. High-aspect-ratio wings are more efficient because a wing becomes less efficient at producing lift as the chord increases. In general, high-aspect-ratio wings have less induced drag but higher parasite drag. A high-aspect-ratio wing can use a smaller engine and needs less takeoff and landing distance than a low-aspect-ratio wing. However, there are many low-speed aircraft with low-aspect-ratio wings, such as some acrobatic airplanes. In some cases, a low aspect ratio and a large engine may be selected to increase the top speed.

Structural considerations also influence the choice of a smaller-aspect-ratio wing. A long, slender wing requires more material and flexes more under the same load. A lower-aspect-ratio wing may save enough on structural weight to offset the reduced lift efficiency because induced power goes as the load squared. Since weight is probably the singularly most important design criterion, the final wing design must take weight into account. A lighter wing means a lighter airplane, and this allows the designer to use a smaller wing area. The smaller wing weighs even less, so there is a cascading positive effect of designing a lighter wing.

> The all-African-American squadron from Tuskegee had trouble convincing superiors that they were qualified for battle. When they finally were sent to Europe, they distinguished themselves by not allowing a single bomber to be lost under their escort.

Another reason to choose a low-aspect-ratio wing, despite its inefficiency, may be for maneuverability. A high-aspect-ratio wing will not roll as quickly as a low-aspect-ratio wing. In aerobatics, the difference in roll rate can be the edge in a competition. The same is true for fighter aircraft. In close combat, it is desirable to have a more maneuverable plane than the adversary.

In contrast, airplanes that are designed for high-altitude flight, such as the U-2 reconnaissance airplane, have very high-aspect-ratio wings, as can be seen in Figure 2.4. At high altitudes, the amount of air available to divert is very small, so airplanes must fly at a high angle of attack. Chapter 1 indicated that high angles of attack mean greater induced power. High-aspect-ratio wings reduce induced power and are the wing of choice for high altitude. A notable exception is the SR-71 Blackbird shown in Figure 2.5. However, that airplane has other

FIGURE 2.4   A U-2 with a high-aspect-ratio wing. (Photograph courtesy of the U.S. Air Force.)

FIGURE 2.5 The SR-71B Trainer. (Photograph courtesy of NASA.)

performance criteria that set it apart from other high-altitude airplanes. This exception will be discussed in Chapter 5.

Sweep

Most modern aircraft use swept wings, as shown in the picture of an X-5 with variable-sweep wings (Figure 2.6). The X-5 had sweep angles of 20, 40, and 60 degrees, with a large jack-screw assembly that could be adjusted in flight. Two were build as test aircraft in the early 1950s.

The primary motivation behind swept wings is to reduce drag at higher cruise speeds. In the late 1930s, it was discovered in Germany that at high speeds the parasitic drag of the wing was related to the angle the air makes with the wing's leading edge. Thus, by sweeping the wing, the drag at high speeds is reduced. This will be covered in greater detail in Chapter 5. In flight near or above the speed of sound, swept wings are mandatory to reduce the power required to sustain

> The horizontal stabilizer on the Boeing 777 has approximately the same area as the wing on a Boeing 737.

ASPECT RATIO OF BIRD WINGS

Birds have different aspect ratios to reflect their needs. Table 2.1 gives some examples of the wing aspect ratio of some birds. In general, birds that fly short distances or must be maneuverable have low-aspect-ratio wings. Birds that travel long distances, particularly over water, have high-aspect-ratio wings, with the wandering albatross an extreme example.

TABLE 2.2 Wing Aspect Ratios for an Assortment of Birds

Bird	Aspect Ratio
Common crow	5
House sparrow	6
Goshawk	6
Golden eagle	8
Herring gull	10
Common tern	12
Wandering albatross	19

FIGURE 2.6 Variable-sweep X-5 research aircraft. (Photograph courtesy of NASA.)

cruise speeds. Most commercial transports, military aircraft, and newer business jets fly between 70 and 90 percent of the speed of sound. Thus they require swept wings. A glance at the wing of a jet tells how fast it is designed to go.

The sweep of the wing also affects stability. A swept wing is generally more stable than a wing without sweep. The mechanism for this stability is quite straightforward. The total drag of a wing can be looked at as acting on a single point on the wing. These are marked as "Centers of Drag" in Figure 2.7. The drag acts on the airplane with equal lever arms on both sides when the airplane is flying straight. If the airplane were to yaw, the drag of the forward wing has a greater lever arm than that of the other wing, tending to realign the airplane with the direction or flight.

> Lieutenant John A. MacReady of the U.S. Army Air Service was attempting a new world altitude record in 1926. He ran out of fuel at 37,000 feet and ended up setting a world record for gliding distance instead.

The enhanced stability owing to backward-swept wings is desirable for passenger airplanes because the airplane will have a tendency to stabilize after it is upset by a gust of air. Conversely, a forward-swept wing will be less stable. Experimentation with forward-swept wings has resulted in less stable airplanes, which generally increases maneuverability.

In the late 1970s, NASA started a program to study swept-forward wings. The program resulted in the X-29 forward-swept wing demonstrator shown in Figure 2.8. The purpose behind the forward-swept wing was to create a naturally unstable platform for enhanced maneuverability. The airplane was so unstable that the pilot alone could not

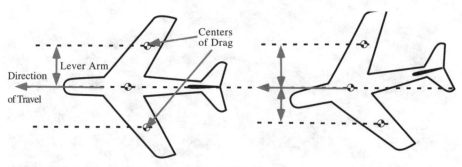

FIGURE 2.7 Effect of wing sweep on stability.

FIGURE 2.8 The X-29 with its forward-swept wing. (Photograph courtesy of NASA.)

control it. A sophisticated computer control system called *fly by wire* was added into the control loop to make the airplane controllable. If the computer were to fail, the airplane would instantly become unflyable.

Sailplanes sometimes have forward-swept wings in order to put the wing spar behind the pilot but with a more forward center of lift. The instability this causes is compensated with increased dihedral. This will be discussed below.

Virtually all aircraft that have swept wings have backward-swept wings. It is worth noting that the high-speed drag reduction owing to sweep is equal for both forward- and backward-swept wings. The primary advantage to forward sweep is to increase maneuverability. However, forward sweep is very difficult to build structurally. A problem known as *structural divergence* can occur if the wing is not stiff enough. What happens is that the forward wingtip twists owing to the load. The twisting increases the tip loading, thus twisting the wingtip even more. Eventually, the tip load is so great that the tip literally twists off the airplane. The advent of modern composites that can be tailored to specific stiffness requirements has led to the possibility of forward-swept wings. Composites can be manufactured in such a way that they can eliminate structural divergence.

Taper

The designer may consider a tapered wing for an airplane. A tapered wing has a shorter chord at the tip than at the root, as shown on the DC-8 in Figure 2.9. There are several advantages to tapering a wing. One reason to taper a wing is to adjust the load along the wing's span. It is desirable to distribute the wing loading such that it is reduced at the wingtip. A high tip loading will put a large bending load on the wing. This means that the entire wing structure has to be built stronger and thus heavier. A rectangular wing will have a large *tip loading* and so require a stronger structure.

> On September 30, 1913, French aviator Roland Garros crossed the Mediterranean in 7 hours and 53 minutes. He had 8 hours of fuel on board.

The primary disadvantage of a tapered wing is that it is more difficult to build. Most small, inexpensive aircraft use at least partial constant-chord wings. The wing of the Cessna 172, shown in Figure 2.10, is a good example of a partial constant-chord wing. A well-known example of a constant-chord wing is the 1960's line of Piper airplanes called the Cherokee, shown in Figure 2.11. These early Cherokees had a wing that was dubbed the "Hershey bar wing" because it is shaped like a Hershey's chocolate bar.

There is another reason to taper a wing, and that is to adjust the lift along the wing to minimize the drag. As discussed in Chapter 1, the lift

FIGURE 2.9   The DC-8 illustrating wing taper. (Photograph courtesy of NASA.)

FIGURE 2.10   Cessna 172. (Photograph courtesy of Cessna Aircraft Co.)

FIGURE 2.11   Piper Cherokee 140 with a "Hershey bar wing." (Photograph courtesy of Albert Dyer.)

When the United States entered World War I, it had 304 airplanes between the U.S. Army Signal Corps and the Marines. Meanwhile, Britain's Royal Flying Corps lost 1270 airplanes in 3 months.

on a wing is proportional to the amount of air diverted by the wing times the vertical velocity of that air. The amount of air diverted, and thus the lift, changes along the span. In classical aerodynamics, it can be shown that the lift of the most efficient wing tapers from the root to the tip in an elliptical fashion. It also can be shown that keeping the downwash velocity, and therefore the angle of attack of the wing, constant along the trailing edge minimizes the induced drag. Thus the elliptical lift distribution is obtained by shaping the wing so that the amount of air diverted tapers in an elliptical fashion. This theory became widely known to designers in the 1930s and resulted in many aircraft with elliptical wings. Two notable World War II–era fighters with an elliptical wing planform are the British Supermarine Spitfire and the Republic P-47 Thunderbolt shown in Figure 2.12.

Figure 2.13 shows the wing loading for a rectangular wing, an elliptical wing, and a linearly tapered wing. The rectangular wing has the highest loading at the tip, whereas the linear taper *unloads* the tip. Today, one does not see elliptical wings on airplanes in part because they are expensive to build. On high-speed airplanes, however, such as commercial transports, when everything is taken into account, such

FIGURE 2.12 The Republic P-47 Thunderbolt with an elliptical wing shape.

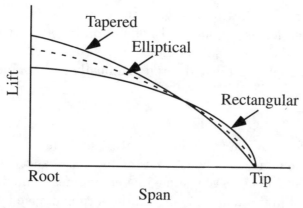

FIGURE 2.13   **Effect of wing shape on the distribution of lift.**

as sweep, dihedral, structural weight, and flight speed, a more triangular distribution is preferred.

Twist

One method of tailoring the lift distribution on a wing is to twist the wing, with the angle of attack greater at the root than at the tip. Another term for this type of twist is *washout*. The lower angle of attack at the tip unloads the tip. There is another advantage to using washout. Because the wing root is at a higher angle of attack, the wing will stall at the root first. Since the ailerons, which control roll, are usually on the outboard portion of the wing, the ailerons still can be effective after the root stalls. If the tip stalled first, the pilot would lose roll control during the stall. This could lead to an uncontrolled stall spin. By designing wings such that the root stalls first, the pilot can control the airplane and prevent the spin.

> The acceleration of air over the top of a wing is the result of the lowered pressure and not the cause of the lowered pressure.

Although mechanically twisting the wing is common on general-aviation aircraft, commercial transports and other high-performance airplanes use an *aerodynamic twist*. Aerodynamics twist results when the wing sections, or airfoils, are changed from root to tip. In other words, different wing-section designs are selected for different positions along the span of the wing. For example, a designer might reduce the camber of the wing sections from root to tip. The goal is to select a wing that is more lightly loaded at the tip and will not stall at the same angle that the wing root stalls. The result is that the wing behaves as though it is twisted, although it is twisted aerodynamically and not mechanically.

Wing Configuration

There is still more to a wing than its airfoil and planform. The configuration of the wing, as viewed from the front, can affect stability, efficiency, and practicality. Why are wings sometimes slanted up or down from the root to the wingtip? Should the airplane have a low wing, a high wing, or a middle wing? What type of wingtip is best? When should one consider a biplane?

Dihedral

Roll and yaw stability are desirable characteristics for trainers and transports, both of which can be enhanced by adding dihedral to the wings. *Dihedral* is the upward angle of the wing along the span against the horizon, as shown in Figure 2.14. Many pilots are taught that the reason dihedral adds stability is that as the airplane rolls, gravity pulls the upward wing back to horizontal. Unfortunately, the truth is more complex.

When an airplane enters a roll, there is a tendency to yaw in the opposite direction. For example, if the airplane rolls to the right, it is accompanied by a yaw to the left. That is, the nose of the airplane swings to the left. This is called *adverse yaw*. This is why the pilot must use compensating rudder to make a *coordinated turn*. The reason for adverse yaw is that the wing rotating upward experiences more drag than the wing rotating down. Whether the rotation was caused by

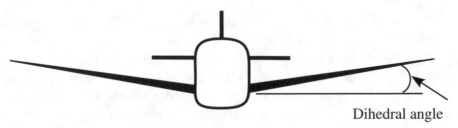

Dihedral angle

FIGURE 2.14   Dihedral angle.

the ailerons or by a gust of air, the airplane rotates because one wing has more lift than the other. We know that lift requires work, and this reflects itself in increased induced drag. Thus the wing with the greatest lift has the greatest drag.

The Wright brothers' number 3 glider of 1902 had no rudder initially. During longer gliding flights, the brothers practiced turns. Much to their dismay, when they banked the airplane, the result was that the airplane went into a *skid* because the nose turned opposite to the bank. They had discovered adverse yaw. The quick-minded brothers deduced the need for a vertical stabilizer and rudder and thus completed the three-axis control puzzle that made controlled flight possible.

Dihedral adds stability because the adverse yaw results in a net reduced angle of attack on the upper rotated wing and an increased angle of attack on the lower (forward) wing. This result is an increase in lift of the lower wing and a reduction in lift of the upper wing, as well as a restoring yaw force. This tends to level the wings and cancel the yaw. To illustrate this, refer to Figure 2.15. Imagine the book on the left is an airplane approaching you and that you see what the oncoming air "sees." That is, both "wings" look the same, and there is a positive angle of attack because the bottoms of the wings are visible. The image of the book on the right is of an airplane that has yawed to the airplane's right. Now the air "sees" the top of the right wing, which is the wing that is higher and has swung back. The air "sees" more of the

FIGURE 2.15   Illustration of dihedral.

bottom of the left wing, which is low and forward. The angle of attack has decreased on the upper wing and increased on the lower wing. This raises the left wing, leveling the airplane, and the adverse yaw realigns the airplane.

Small general-aviation aircraft and commercial transports all have dihedral. These airplanes tend to return to level flight after gusts or accidental control inputs. Today, airplanes are so stable that cross-country trips can be quite boring.

> During the early days of flight, the Wright brothers advocated a design in which the airplane was slightly unstable. They felt that this forced the pilot to pay attention and continually fly the airplane. Meanwhile, competitors built airplanes that were inherently stable. Eventually, the Wright brothers could not compete with these stable designs. This is one factor that has led to the eventual loss of the Wright name in any airplane company today.

Some military aircraft, such as the F-104 (Figure 2.16), have anhedral wings. *Anhedral* is negative dihedral; in other words, the

FIGURE 2.16 F-104 with anhedral (negative dihedral). (Photograph courtesy of NASA.)

wings slope down, as seen in the figure. Just as dihedral is stabilizing, anhedral is destabilizing. As mentioned earlier, stability tends to reduce maneuverability. By making the airplane less stable, it becomes more maneuverable. Anhedral is found mostly on military fighters and aerobatic airplanes.

For completeness, we should mention another stabilizing effect of all wings. Consider an airplane that is rotating because of a gust of air or a sudden short movement of the ailerons. After the gust of air is over or the ailerons are back to neutral, the lift on both wings would seem to be the same, although the inertia keeps the airplane rotating. However, while the airplane is in rotational motion, the wing going up "sees" the relative wind with a reduced angle of attack, and the lower wing "sees" an increased angle of attack. This causes a force opposite to the rotation and tends to bring the rotation to a stop. When the rotation stops, the counterrotational force also ceases.

> A popular home-built airplane has had problems flying in rain. The Vari-EZ and Long-EZ, shown in Figure 2.20, have laminar-flow canards. When flown in rain, the water droplets ruined the laminar-flow properties sufficiently for pilots to notice changes in handling characteristics.

High Wings vs. Low Wings

The wing designer also must decide where to put the wing on the airplane. Should the wings be low, high, or in the middle of the fuselage? What are the benefits of wing position? The effects of wing position on stability are fairly minor. A high wing has a little more stability than a low wing, but the change in stability is small compared with the effects of dihedral and wing sweep. Thus the choice of wing position is based on much more practical matters. Private pilots frequently argue this point in hangar talks. Those who fly low-wing airplanes insist that the clear view of the sky outweighs the lack of visibility downward because of the wing. The high-wing pilots argue that the view of the ground is more important than the view of the sky. Those who fly in hot, sunny regions also appreciate the high wing for shading the cabin.

Besides a slight increase in stability, a high-wing aircraft offers the ability to locate the fuselage close to the ground. Military transports use this configuration so that equipment can be loaded and off-loaded easily. A good example is the C-130 Hercules shown in Figure 2.17.

FIGURE 2.17   C-130 Hercules demonstrating fuselage upsweep. (Photograph courtesy of the U.S. Air Force.)

High wings also offer more room for high-lift devices. For example, wing flaps can extend further down without a concern for ground interference. Another advantage is that wing struts can be used under the wing, where they will not interfere with lift.

A disadvantage of high wings is that the landing gear must be placed in the fuselage. This usually adds bulging pods to accommodate the gear. Another disadvantage is that the low fuselage leaves little tail clearance. In order for the airplane to be able to rotate at takeoff, the fuselage must have upsweep, as shown in Figure 2.17. This sacrifices valuable cargo space.

Low-wing airplanes make landing gear placement much simpler. However, for multiengine airplanes with engines hung on the wings, the landing gear must be long enough to prevent ground interference with the engine. Low-wing airplanes also have structural benefits because the wing *spars* (the internal beams supporting the wing) can carry through the lower fuselage below the passenger deck, resulting in a continuous spar structure. This allows the wing to be fully *cantilevered* (supported by one end) with no need for external bracing. Most commercial transports since the DC-3 have chosen this approach. Because the fuselage is positioned higher than for a high-wing airplane, there is less need for fuselage upsweep. The primary disadvantage is the requirement for long landing gear to provide ground clearance for the engines.

Airplanes with wings positioned in the middle of the fuselage are usually only found in military fighter or aerobatic airplanes. The mid-position offers the benefit of clearance for stores underneath the wing

COWLING SAVES THE BOEING 737

The Boeing 737 was designed initially with turbojet engines (see Chapter 4). To improve efficiency, a move was made to turbofan engines, which have a bigger diameter. The Boeing 737 had been designed close to the ground because the small-diameter turbojets did not need much clearance. Boeing had difficulties retrofitting the Boeing 737 with turbofan engines with so little clearance. The result is an unusual cowling design (Figure 2.18) that is flat near the bottom. The inset in the picture is of the old engine for comparison. Creating this innovative cowling saved the Boeing 737, which now has become the most popular commercial jet airplane in history.

FIGURE 2.18 The Boeing 737 noncircular engine inlets, with the old engine shown in the inset.

while maintaining visibility above. The configuration also has the lowest drag. The primary disadvantage of a midwing configuration is that the wing-fuselage joint occurs in the middle of the fuselage. The structure must carry through the fuselage at a point where one might want to put passengers, cargo, etc.

Wingtip Designs

The design of the wingtip has an effect on lift. A nice rounded wingtip looks clean but actually results in lower performance than a square tip. The design of the wingtip also affects the distribution of lift along a wing and thus the wing's efficiency. Thus the shape of the wingtip has become an important design criterion.

Choosing the correct wingtip is a matter of compromises, which include aerodynamic performance, structural loads, manufacturing, and (possibly most important) marketing. In the early days, rounded tips were simple to build because all the builder had to do was bend a rod from the leading edge around to the trailing edge. Since loads are small at the wingtip, little internal structure was needed. Many of the pre–World War II airplanes had rounded wingtips, but squared-off tips are aerodynamically more efficient. The squared-off wingtip better restricts the passage of high-pressure air from the lower surface to the upper surface. Any high-pressure air leaking to the upper surface leads to lower aerodynamic efficiency. Thus most of today's airplanes have simple squared-off tips.

> The wing of a Boeing 747 has the same parasitic drag as a ½-in cable of the same length.

Winglets

Today, many aircraft sport *winglets*, which are wingtips turned vertically, as shown in Figure 2.19. While the winglets usually point up on wings, it should be noted that winglets on horizontal stabilizers point down because the horizontal stabilizer normally pulls down, as will be discussed in Chapter 3. Winglets go one step further in preventing the passage of high-pressure air from flowing around from the lower surface to the upper surface. In essence, they provide a block. The result is that the wing can carry a finite amount of lift all the way to the tip. As discussed in Chapter 1, the efficiency of a wing increases with size. Winglets increase the effective length of the wing and thus

FIGURE 2.19   A winglet on a K-C135A. (Photograph courtesy of NASA.)

increase the wing's efficiency without increasing its length. Winglets have become the golden child of all airplane sales representatives.

However, winglets do come with disadvantages. The net effect of the winglet is at best equivalent to laying the winglet flat, which would result in extra span and wing area. Thus one could achieve the same wing performance by doing exactly that, laying the winglet flat. The added tip loading owing to the winglet requires a stronger and heavier wing (but not as much as adding span). Therefore, winglets cannot be retrofitted to existing airplanes without changing operating conditions to lower the wing loading or strengthening the wing. On a new design, though, winglets can help to make the wing more efficient by reducing the induced power required for a given lift, just as extra span would. Perhaps the best benefit is that greater efficiency can be achieved without increasing span if there is a span constraint, such as a hangar or airport

> During World War II, the United States was producing 5500 aircraft per month.

gate restriction. Nevertheless, one of the biggest reasons for the preponderance of winglets on today's business jets is that they are considered "very sexy."

One might wonder why the winglets of some of the jets cover only the last part of the wingtip, as shown in Figure 2.19. This is because, at cruise speeds, the air flowing over the first part of the wing is going at or near the speed of sound. This will be discussed in Chapter 5. A winglet on the forward part of the wing of a transonic jet could cause a shock wave and increased drag.

Canards

The Wright brothers designed their airplanes such that the horizontal stabilizer was forward of the wing. This configuration is called a *canard*, which is the French word for duck. However, the canards soon disappeared from the scene, until their recent revival. Today, canards have become popular, much like wingtips, because of their "sexy appearance." A dramatic example of a canard is the NASA X-29 experimental aircraft shown in Figure 2.8 and the Long-EZ shown in Figure 2.20.

Canards have the advantage that the horizontal stabilizer is lifting up rather than down, as is typical on a conventional airplane. This reduces the load on the wing. Thus the canard appears to be more efficient because the wings and the horizontal stabilizers provide lift. However, canards must be designed such that the horizontal stabilizer

FIGURE 2.20 A home-built Long-EZ illustrating a canard configuration. (Photograph courtesy of Sandy DiFazio.)

stalls before the wing. If the wing stalled first in the canard configuration, the rear of the airplane would drop, increasing the angle of attack further, and stall recovery would be impossible. The canard is designed such that the horizontal stabilizer stalls first, dropping the nose of the airplane. Thus the airplane's wing will not stall. This safety feature has encouraged many modern designers to favor canards.

> The Messerschmitt BF-109 was the most-produced fighter aircraft in history, with about 35,000 built.

Canards are touted as being more efficient than conventional airplanes, but this conclusion ignores induced-power requirements. The horizontal stabilizer is at a higher angle of attack, so it is working harder than the wing. As much as one-third of the airplane's weight may be supported by the canard alone. By comparison, the wing is loafing. For efficiency, it would be better to have the wing working harder. Thus the canard sacrifices efficiency so that the horizontal stabilizer always stalls first.

Boundary Layer

The concept of the boundary layer was introduced in Chapter 1. In order to understand certain aspects of wing design, it is necessary to go into more detail. This section will discuss the effects of energy on the boundary layer as an introduction to high-lift devices.

The boundary layer on a wing section is shown in Figure 2.21. In this figure the boundary-layer thickness is greatly exaggerated. In reality, the boundary layer is quite thin. For example, at the trailing edge of a Boeing 747-400 (where the boundary layer is the thickest), the boundary layer is approximately 6 in (15 cm) thick. This is close to the root of the wing, where the chord is about 30 to 35 ft (10 to 12 m). Figure 2.22 is a blowup of the boundary layer showing how the speed relative to the wing changes from zero relative to the wing from the wing surface outward. It is the friction of air molecules with the surface and viscosity that causes this change in speed.

The lower speed near the surface translates into lower kinetic energy. Boundary-layer energy is important because higher energy will allow the boundary layer to continue to follow a surface even as the pressure increases gradually. As stated in Chapter 1, most of the lift on a low-speed wing is in the first one-quarter of the chord length. This is

FIGURE 2.21 The boundary layer.

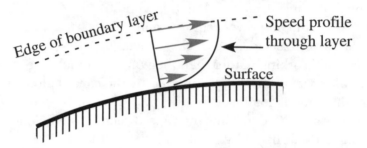

FIGURE 2.22 How the air speed changes in the boundary layer.

where the pressure is lowest. As the air moves back along the top of the wing, the pressure increases until it reaches the ambient pressure at the trailing edge of the wing. This is known as the *trailing-edge condition.* Thus the air is moving into an increasing pressure, which slows it down. This is very similar to a ball rolling uphill. If the boundary layer has enough energy to overcome the increasing pressure, it will follow the wing's surface. When the energy of the boundary layer is not sufficient, the boundary layer will stop flowing and separate from the surface. Past the separation point, the wing experiences air flowing in the reverse direction. The wing is entering a stall. Separation usually occurs near the trailing edge at the critical angle of attack. As the angle of attack increases, the point of separation moves forward, the lift decreases, and the drag increases.

An understanding of the energy in the boundary-layer air is necessary for wing designers to design wings that hinder separation. If a wing can reach a higher angle of attack before stalling, it will be able to take off and land at lower speeds or carry a heavier load. Lower stall speeds translate into shorter runways, and heavier loads translate into greater revenue.

ICE ON THE WING

A wing designed to stall from the trailing edge first may lose this characteristic when it flies into icing conditions. Ice forms on airplanes that fly into moisture in a certain temperature range. Supercooled water drops freeze on impact and form *rime ice,* which is rough and opaque. Water that is warmer will hit the wing and form *glaze ice,* which is smooth and clear. *Mixed ice* is a mixture of glaze and rime ice.

The buildup of ice does several negative things to the wing. The first is that it changes the shape of the wing and thus changes its stall characteristics. In general, a wing with ice will stall abruptly at a lower angle of attack than without ice. Ice also adds weight, which is an additional load on the wing. Therefore, to compensate, the pilot must increase the angle of attack. Eventually, the angle of attack required to maintain flight reaches the new stall angle of attack, and the airplane can no longer fly. An additional impact of ice is increased drag, both induced and parasitic. The result is that the power requirement increases.

Boundary-Layer Turbulence

Normally, as passengers or pilots, we associate turbulence with atmospheric conditions. This is known as *clear-air turbulence.* Clear-air turbulence is important to structural designers to design for gust loading. However, there is another type of turbulence that is critical to the aerodynamic design of the wing. This is *boundary-layer turbulence,* which occurs only in the very thin boundary layer.

Most pictures of wings with air flowing over them show smooth patterns of air. This is *laminar* flow, which is nonturbulent flow. Figure 2.23 illustrates both laminar and turbulent flow. In the figure, smoke in laminar-flow air passes through a special screen. Shortly after the screen, the laminar flow becomes turbulent. The basic difference is that the laminar flow is smooth, whereas the turbulent flow is chaotic. Skin friction and skin drag increase dramatically with the increase in boundary-layer turbulence. A wing with laminar flow will have much less skin drag than a wing with turbulent flow.

It would be a great asset to design a wing that was laminar over its entire surface. But this has proven extremely difficult. Boundary-layer

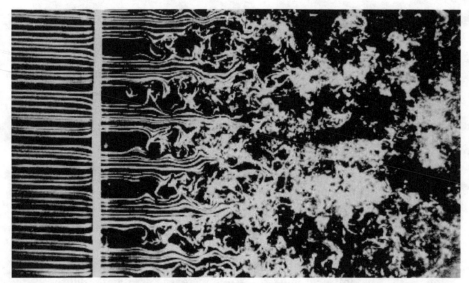

FIGURE 2.23 Laminar flow turning into turbulent flow. (Photograph courtesy of Thomas Corke and Hassan Nagib.)

turbulence is a natural phenomenon, and it becomes more prevalent as speed and the size of the airplane become greater. Besides this natural tendency of a laminar boundary layer to become turbulent, things such as rivets, bugs, and raindrops can trigger boundary-layer turbulence. A great deal of effort has gone into understanding this turbulence and devising ways to reduce or eliminate it. Modern wings attempt to maintain laminar flow as far back on the wing as possible. It then transitions to turbulent flow before the air has a chance to separate. If you look at the wing of a commercial jet, you will see that it is smooth part of the way back, and then there is less concern for smoothness further aft.

Turbulence does have one beneficial effect. Because the airflow in the boundary layer is churning, it mixes the more energetic air with air close to the wing surface. Thus the boundary layer is "energized" and has increased the kinetic energy near the surface. The advantage of this is that the air will stay attached to the surface longer. A laminar-flow wing will stall at a lower angle of attack than a non-laminar-flow wing. Another method for energizing the boundary layer is to add *vortex generators*. Many airplanes use vortex generators to delay separation of the air and thus delay stall. Vortex generators will be discussed below.

Form Drag

We should look at drag a little more carefully to better understand stalls. Parasite drag is composed of two parts. Chapter 1 discussed the effect of friction. Also mentioned was form drag. Form drag is the drag associated with moving things such as antennas and wheels through the air. Form drag can be thought of as the drag associated with pulling the wake of the airplane along with it. Form drag plus skin drag make up parasite drag.

Form drag is what people usually associate with aerodynamic shapes. People often look at cars and instinctively know which are better aerodynamically. We tend to look at how streamlined the car is. What usually distinguishes a streamlined car from an unstreamlined car is the apparent wake resulting from the rear of the car. A truck clearly has a large wake and high form drag, as illustrated in Figure 2.24. It is desirable to reduce form drag on an airplane as much as possible. This means reducing the wake. Any part of an airplane where the air separates from the surface produces a wake. Even a small, cylindrical antenna will produce a significant wake. Therefore, protuberances such as antennas are encased in aerodynamic fairings.

> Charles Lindbergh flew 50 combat missions during World War II, even though he was a civilian.

Until about 1920, airplane designers thought that wings had to be thin. To create an efficient structure, there was much use of external wire bracing. Ironically, the external bracing resulted in much higher drag than what was saved by making the wings thin. Late in World War I, the Germans discovered that the wire bracing was adding too much drag and started to use fatter wings that could hold more structure.

FIGURE 2.24   Form drag illustrated by the wake of a truck.

In discussing the stall of a wing, we have focused on the reduction in lift. Almost as great a problem is the wake caused by a stalling wing producing significant form drag. Just when the airplane needs all the power it can get to recover from the stall, there is a large additional power requirement put on the airplane, making recovery even more difficult.

Vortex Generators

A boundary layer with more kinetic energy will be less prone to separate. *Vortex generators* are a means for adding kinetic energy to the boundary layer and thus increasing the stall angle of attack. Vortex generators are simple devices that can be retrofitted to any airplane. They are simply small, flat plates perpendicular to the wing's surface set at an angle of attack to the local air, as shown in Figure 2.25. Vortex generators add kinetic energy to the boundary layer by mixing in higher-kinetic-energy air from above. For vortex generators to be affective, they only need to be as high as the boundary layer is thick. They are usually added a little aft of the region of highest air velocity. On the airplane shown in Figure 2.25, the vortex generators are also added to the leading-edge slat, which is a high-lift device discussed below.

Unfortunately, there is a penalty for adding vortex generators. Since vortex generators put a force on the air, they do work, yet without adding to the lift to the wing. Therefore, additional power is

> Orville Wright made a 9-minute flight in a glider. It remained a world record for 10 years.

FIGURE 2.25   Vortex generators on the slat and wing of a Navy A-4 Sky Hawk.

needed when vortex generators are used. The reality of vortex generators is that they are used most often to correct an existing problem.
Because they introduce a power penalty, there is no incentive for
putting them on the original wing design. A well-designed wing
should not need vortex generators.

Nevertheless, there are several cases where
it is desirable to use vortex generators. The first
is in situations where they are used only for
takeoff and landing but are stowed for cruise. A
particular situation where this can be seen is on

> In 1919, Winston Churchill was
> named Britain's air minister.

the leading edges of some wing flaps. The use of flaps will be discussed in the next section.

The second case where vortex generators are desirable is when an
existing airplane is modified to meet new performance parameters. A
typical change in performance requirements is to make an existing airplane use less runway for takeoff and landing. Ideally, a new wing
would be designed. However, designing a new wing for an existing airplane is usually too costly. Vortex generators thus are added to
increase takeoff and landing performance. There are many companies
that modify existing airplanes, converting them to STOL airplanes.

High-Lift Devices

Devices or modifications to the wing that increase the stall angle of
attack are called *high-lift devices*. Airplanes employ various high-lift
devices to improve takeoff and landing performance. The basic principle behind these devices is that they allow the wing to divert more air
down without stalling. Vortex generators, discussed above, are a type
of high-lift device. During takeoff and landing, the airplane needs to
fly at the lowest speed possible because high takeoff and landing
speeds mean longer runways. Therefore, a major goal in designing a
wing is to reduce the stall speed as much as possible. The easiest way
to do this would be to design the wing with a great deal of wing area
and camber. Such a wing would be able to fly at a lower speed without
stalling. As mentioned earlier, however, a high-cambered wing would
have a high drag at cruise speeds. The solution is to design a wing that
can change its characteristics for takeoff and landing speeds. A second

THE GOLF BALL

Turbulence can have a beneficial effect on reducing the wake of an object. If the surface air in the boundary layer becomes turbulent, the higher kinetic energy in the turbulent flow will help the air stick to the surface longer before separating. The result is lower form drag.

A smooth ball traveling through the air will have high form drag, as shown in Figure 2.26. The laminar flow soon separates, producing a wake behind the ball that far outweighs the skin drag. One solution to improve the range of a ball is to energize the air around the ball by churning up the air. This will allow the air to remain attached to the surface longer and reduce the size of the wake.

A golf ball's size normally would result in laminar flow and a sizable wake. This is why the surface is covered with dimples, which encourage turbulence. As seen in the figure, the dimples reduce the size of the wake and thus reduce the form drag. The result is that golf balls with dimples travel much farther than if they were smooth.

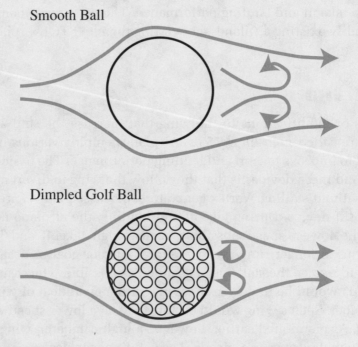

Smooth Ball

Dimpled Golf Ball

FIGURE 2.26   The dimples on a golf ball decrease form drag.

way to increase the stall angle of attack, discussed earlier, is to manipulate the boundary layer.

The most common high-lift device is the wing flap. The next most common is to add leading-edge devices called *slots* and *slats*. In rare instances, the deflected slipstream from the propellers or jet engine is diverted to provide additional lift at low speeds.

Flaps

Wing flaps can be found on virtually every modern airplane. The effect of adding flaps to the trailing edge of the wing is equivalent to increasing the camber of the wing. Some flap designs also increase the chord length of the wing. This increases the area of the wing so that more air is diverted, thus reducing the angle of attack needed for lift.

There are many types of flaps. In the 1930s and 1940s, the *split flap*, shown in Figure 2.27, was introduced and was one of the first types of flaps to appear in production airplanes. Splitting the last 20 percent or so of the wing forms this type of flap. The top surface of the wing does not move, whereas the bottom surface lowers. The split flap is effective in improving the lift, but it creates a great deal of form drag, as shown in the figure. The split flap was used on the DC-3 as well as on the Boeing B-17, as shown in Figure 2.28. It also was used on World War II–era dive-bombers because it helped to increase lift at low speeds and slowed the airplane during the dive.

On February 21, 1979, at Kitty Hawk, former astronaut Neil Armstrong climbed to 50,000 ft in a business jet and set five world records.

The simple *hinged flap* (Figure 2.29) is most common on smaller aircraft. The last 20 percent or so of the inboard section of the wing is simply hinged so that it can increase the camber. The first 20 degrees of flap extension increase the lift without greatly increasing the drag of the wing at low speeds. Many airplanes extend their flaps to 10 or 20 degrees on takeoff in order to shorten the takeoff distance. When the flaps are extended greater than 20 degrees, the form drag increases rapidly with little or no increase in lift. Increasing the drag increases the descent rate, which is desirable during the approach for landing. Thus it is not uncommon for an airplane to land with the flaps set at 40 degrees.

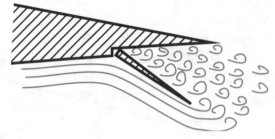

FIGURE 2.27 Split flap.

A more sophisticated flap is the *Fowler flap* shown in Figure 2.30. With the Fowler flap, the rear section of the wing not only changes angle but also moves aft. The result is both an increase in camber and an increase in wing area. A bigger wing will divert more air, and increased camber will increase the downwash velocity. Mechanisms to operate Fowler flaps can be quite complicated. This type of flap is common on smaller Cessna airplanes such as the 172 and 182.

FIGURE 2.28 Split flap on a B-17 bomber.

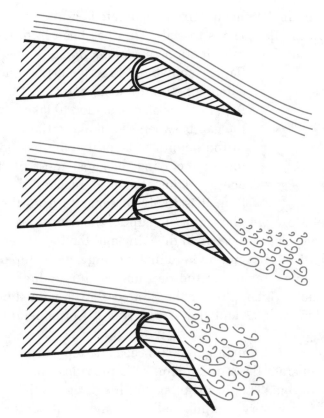

FIGURE 2.29 Hinged flap.

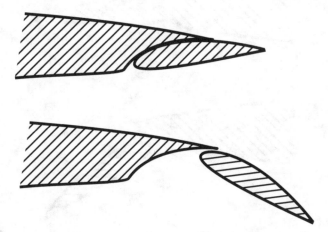

FIGURE 2.30 Fowler flap.

The maximum lift that a flap can generate is limited by the critical angle where the flap begins to stall. This has been improved by the introduction of *slotted flaps*. A single slotted flap is shown in Figure 2.31. A slotted flap extends both aft and downward, like a Fowler flap, but it is also designed to take advantage of the gap between the flap and the wing. The air in the boundary layer, having passed over the top of the wing, has lost most of its kinetic energy. Thus, when it reaches the extended flap, it is likely to separate from the flap and cause a stall. However, the air passing under the wing does not face the same problem. The slot between the wing and the flap diverts some of the lower-surface air, with higher kinetic energy, to the top of the flap. The air remains attached to the flap longer, thus reducing drag and inhibiting stalls. Examples of the slotted flap are the Boeing C-17 shown in Figure 2.32 and the World War II German Junkers Ju 52 shown in Figure 2.33.

A *double-slotted flap* (Figure 2.34) basically repeats this step twice, using two separate flaps in tandem. This provides the maximum lift from a flap design. The disadvantage of this design is that the operating mechanism is very complicated and heavy. Multislotted flaps are

> Sometimes technology does not rule. During the Vietnam War, 91 percent of all U.S. fighters shot down by antiaircraft fire were aimed at by hand.

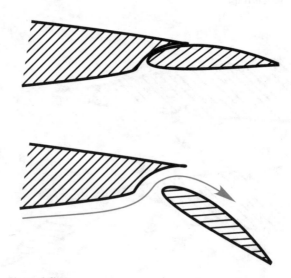

FIGURE 2.31   Slotted flap.

FIGURE 2.32   Slotted flaps on a Boeing C-17.

FIGURE 2.33   Slotted flaps on a World War II German Junkers Ju 52.

seen on many of the modern passenger jets, whereas smaller airplanes use single-slotted flaps. A good example of a triple-slotted flap is shown in Figure 2.35.

Until the 1990s, airplane performance was the key design criterion. Airplane companies were proud of sophisticated triple-slotted flap systems. During the 1990s, a shift toward reducing cost as a key design criterion pushed airplane companies to maximize the performance of single-slotted flaps. One technique that is used is to place vortex gen-

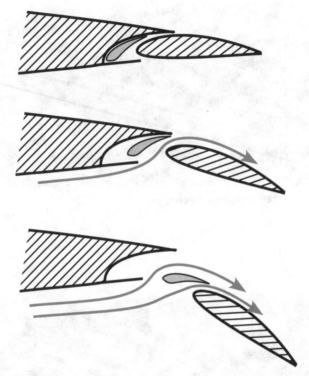

FIGURE 2.34 Double-slotted flap.

FIGURE 2.35 Triple-slotted flaps on a Boeing 747.

erators on the leading edge of the single-slotted flap. When the flap is retracted, the vortex generators on the flap are hidden in the wing. Thus the vortex generators do not penalize the airplane in cruise but are available for takeoff and landing.

The next time you fly a commercial airplane, ask for a window seat behind the wing. During the approach and landing phase of the flight, watch the wing unfold. It is truly remarkable how the wing evolves into a high-lift wing from its normal cruise configuration.

In 1709, Brazilian Jesuit Bartolomeu Lourenco de Gusmao demonstrated the first hot-air balloon to the King of Portugal. Unfortunately, the fire used in the balloon ignited the royal draperies, causing considerable damage.

Slots and Slats

Like flaps on the trailing edge of the wing, leading-edge devices are sometimes used to increase the camber of the wing, but they also increase the stall angle of attack. But the details are somewhat different. These devices allow the air from below the wing to flow to the upper surface of the wing and reduce the magnitude of the low-pressure peak near the leading edge. This reduces the pressure difference between the low-pressure region and the trailing edge. The net backpressure is less, and the air is able to negotiate the upper-wing curvature. Thus the wing stalls at a higher angle of attack, and maximum lift is increased.

The simplest leading-edge device is the *fixed slot* shown in Figure 2.36. This is a permanent slot near the leading edge of the wing. The high-pressure air below the wing is drawn up through the slot and flows over the top of the wing. This "energizes" the boundary on top of the wing at high angles of attack. A permanent slot can increase the critical angle of attack significantly. The disadvantage of a fixed slot is

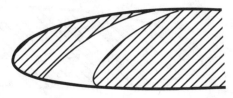

FIGURE 2.36 Fixed slot.

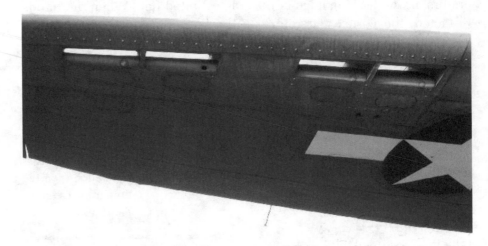

FIGURE 2.37   Fixed slots on the wing of a Lockheed C-60 Lodestar.

that it causes increased power consumption and drag at cruise speeds. Figure 2.37 shows the fixed slots on the wing of a Lockheed C-60 Lodestar.

A similar device to the slot is the *fixed slat*, shown in Figures 2.38 and 2.39. It is added onto the wing, increasing the wing's chord length as well as energizing the boundary layer. Like the fixed slot, the fixed slat causes increased drag at cruise speeds.

> The Boeing 747's wingtips can flex 26 ft (8 m) before they snap off.

The solution to the drag caused by fixed slots and slats is to design a slat that is deployed only at slow speeds and causes little or no drag at cruise speeds. The Handley-Page retractable slat, shown in Figure 2.40, extends to a large droop angle to give the wing large leading-edge camber. In cruise, the slats are retracted and do not cause increased drag. These types of slats are often designed so that they deploy by

FIGURE 2.38   Fixed slat.

FIGURE 2.39   Fixed slat on a STOL airplane.

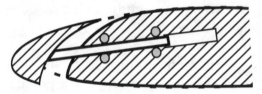

FIGURE 2.40   Handley-Page retractable slat.

themselves at slow speeds and high angles of attack and return to the flush position in cruise.

Deflected Slipstream and Jetwash

One way to increase lift at slow flight speeds is to divert the propeller's slipstream or the jet engine's exhaust down. To achieve a substantial increase in lift with a slipstream, the plane must have engines mounted on the wings with large propellers that generate a slipstream over a substantial portion of the wing. The wing also must have a multislotted flap system to deflect the slipstream effectively. This technique has not found significant commercial application.

A tail dragger is also known as conventional gear because before World War II, nose wheels were rare.

The exhaust of a turbofan-powered airplane can be diverted down to produce additional lift at low speeds. One way to produce the diversion is to have the flaps extend down into the exhaust when fully extended. One problem with

> There are over 500,000 licensed pilots in the United States.

this technique is that the flap extension into the jet exhaust exposes it to very high temperatures, creating a significant design challenge.

Another way to divert the jet exhaust is to mount the engines on the top of the wing with the engine exhaust crossing the top of the wing as in Figure 2.41. The air follows the surface of the wing just as the oncoming air does. Flaps behind the engines divert the exhaust down when extended. This gives a substantial increase in lift for takeoff and landing.

Wrapping It Up

Before moving to Chapter 3, let us look at some typical wings and identify what choices were made in designing them. We will look at the following three airplanes: the Cessna 172 from general aviation, the Boeing 777 from commercial aviation, and the Lockheed-Martin/Boeing F-22 Raptor from the military.

The Cessna 172, shown in Figure 2.10, is a popular four-seat airplane with a cruise speed of 140 mi/h (224 km/h). The wing is unswept but has a small amount of taper toward the tip. The wing is mounted at the top of the fuselage for structural reasons. The wing has an aspect ratio of 7.5, dihedral for stability, Fowler flaps for landing,

FIGURE 2.41 NASA QSRA STOL research vehicle with deflected jetwash. (Photograph courtesy of NASA.)

and round leading edges for a gentle stall. The low wing loading gives it good low-speed performance, but its top speed is not particularly noteworthy. The airplane is designed to be easy to fly and has the best safety record of all general-aviation aircraft.

The Boeing 777 (Figure 2.42) was designed to carry a heavy load over long distances. Its high cruise speed of about Mach 0.84 requires a high wing loading. The aspect ratio is similar to that of the Cessna 172, but the wing is swept and tapered. In this case, the sweep is necessary for the high cruise speed. Taper, as well as changes in the wing section, results in a tailored lift distribution at cruise. For takeoff and landing, the airplane has double-slotted flaps and deployable leading-edge slats. The wing is fairly thick (a person can stand in the wing root) to accommodate the necessary structure and fuel. The wing also has significant dihedral for stability.

> In 1909, the Wright brothers filed suit against Glenn Curtiss for patent infringement. The Wrights contended that Curtiss's use of ailerons was the same as wing warping. The lawsuit held up development of aviation in the United States until World War I. Eventually, the Wrights won, but by that time the consequences were irrelevant.

The Lockheed-Martin/Boeing F-22 Raptor is the current-generation stealth fighter (Figure 2.43). The body and wing blend together for stealth. As a result, it shares the characteristics of a low-wing and a midwing design. The wing is more highly swept and tapered than the Boeing 777. It is highly maneuverable, can cruise at greater than the speed of sound, and is said to have a top speed of more than 1600 mi/h

FIGURE 2.42   Boeing 777. (Used with permission of Boeing Management Company.)

FIGURE 2.43 Lockheed-Martin F-22 stealth fighter. (Photograph courtesy of the U.S. Air Force.)

(2,500 km/h), which is Mach 2.4. The F-22 Raptor employs many high-lift devices, including diverting engine thrust down. Its wing loading is high, and it has a very small aspect ratio.

> The cost of modern airplane computer and electronics systems is approaching 50 percent of the cost of the airplane.

In this chapter we considered the factors that go into design of a wing. We also discussed the stability created by sweep and dihedral. This is called *lateral stability*. It is time to look at stability in the other two axes and to better understand the meaning of stability. In Chapter 3 we will do just that as we look at stability and control.

Stability and Control

One of the greatest improvements in aircraft over the last few decades has been in the area of stability and control. *Stability* is the tendency of an airplane to return to a previous condition if upset by a disturbance such as a gust of air or turbulence. *Control* is the ability to command the airplane to perform a specific maneuver or to maintain or change its conditions. Before World War II, stability and control did not receive much emphasis. The issue was merely how well the pilot thought the airplane "handled." Different philosophies reigned. For example, the Wright brothers felt that a less stable airplane was better because they believed that it forced the pilot to be diligent. The Wrights' competitor, Glenn Curtiss, believed that an airplane should be very stable to reduce the pilot's workload. The Curtiss camp prevailed in the long run.

After World War II, engineers started to develop quantitative means for determining airplane stability and handling properties. Today, the computer can do the work of controlling an unstable airplane while the pilot focuses on other tasks. In these next few sections we guide you through some basic principles of stability and control.

Static Stability

There are three types of static stability. These are statically *stable*, *unstable*, and *neutral*. All three are illustrated in Figure 3.1. The ball in the bowl illustrates a statically stable system. If the ball is displaced from the bottom, it will tend to return. An increase in the steepness of the sides of the bowl corresponds to an increase in stability. If we were to turn the bowl over, we would have an illustration of static instability. If the ball is moved from the exact center of the bowl, it will continue to move away. The ball on the table illustrates neutral stability. If the ball is moved, it will tend to stay in its new position.

For an airplane, static stability means that if a gust of air or some other perturbation causes a change in its current state such as heading, it will experience a restoring force. Small general-aviation and commercial aircraft, if trimmed properly, will return to straight-and-level flight after a wind gust or an abrupt disturbance of the controls. There have been instances where a pilot has fallen asleep only to have the airplane continue its course without an autopilot. Even Lindbergh got a short, unplanned nap on his solo trip across the Atlantic and lived to tell about it.

> John Northrop was a talented engineer who worked for both Lockheed and Douglas. Northrop was a strong advocate of flying wings. He started his own company to develop the flying wing, with the first model a piston-powered version of the YB-49 (Figure 3.4).

Stability should not be confused with balance. An airplane is balanced when there is no net torque on the airplane. Stability is the tendency for the airplane to return to its previous attitude once disturbed. In the next section we lead you through pitch stability, which should make this distinction clear.

Longitudinal Stability and Balance

Longitudinal stability is the tendency for an airplane at a specific pitch attitude to return to that attitude when perturbed. We are interested in the stability of the airplane as a whole.

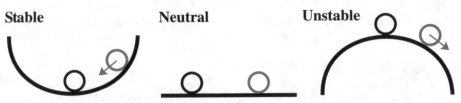

FIGURE 3.1   Three types of static stability.

Balance, on the other hand, occurs when there is no net torque on the airplane. This means that the airplane is not trying to role or pitch or yaw. To be balanced, the center of lift of the entire airplane and the center of gravity must coincide. The *center of lift* is the sum of the lift from the wing combined with the lift of the horizontal stabilizer.

Horizontal Stabilizer

The horizontal stabilizer allows adjustment of the center of lift. If the horizontal stabilizer is lifting upward, it moves the center of lift aft. Picture an airplane in which the horizontal stabilizer had the same lift as the wing. The center of lift would be halfway between the two surfaces. If the horizontal stabilizer is pulling down, it moves the center of lift forward. Thus the horizontal stabilizer can be used to balance the airplane. But how does it contribute to stability? Before we can answer this question, you must understand the concept of the neutral point.

In simple terms, the *neutral point* is the farthest aft that the center of gravity of the airplane can be placed and the airplane still have stable flight. For balanced flight, the center of lift must be at the center of gravity. For stable flight, the center of lift must be at or forward of the neutral point.

In 1955, "Tex" Johnston became a legend at Boeing when he rolled the Boeing 707 prototype at an air show in front of many airline executives.

Figure 3.2 shows a toy airplane in its trim condition (i.e., straight-and-level flight) and then perturbed with an increased angle of attack. Three examples are illustrated: stable, neutrally stable, and unstable. The top example is chosen such that the center of gravity is far forward and ahead of the neutral point. Looking at the picture, we see that the lift of the wing produces a torque that wants to push the nose down because the airplane rotates about its center of gravity. To counter this rotation and thus balance the airplane, the horizontal stabilizer must produce a downward force on the tail of the airplane. Notice how the horizontal stabilizer is at a small negative angle of attack. Because the tail is so far aft, there is a long lever arm so that the downward force on the tail only needs to be on the order of 10 percent or less of that of the wing (typically 6 to 8 percent). Because of the downward force on the tail, the center of lift moves forward from the center of lift of the wing to coincide with the center of gravity. The toy airplane is balanced.

Stable

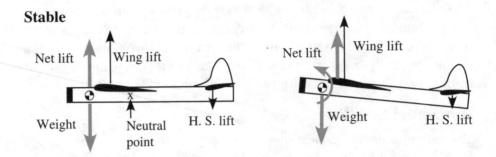

⊙ Center of gravity (c.g.)

Neutral

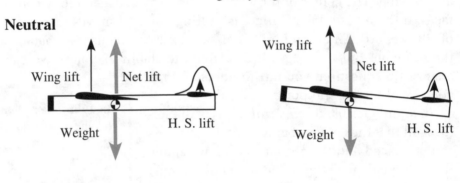

Unstable

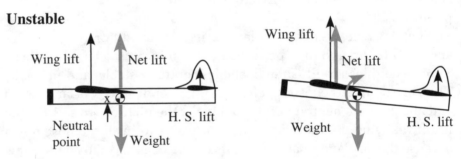

FIGURE 3.2 **Stability of a toy airplane.**

The picture on the top right in Figure 3.2 shows the stable toy airplane perturbed from its trim condition. Just as an example, let us say that a gust of air has increased the angle of attack by 5 degrees. The wing will create more lift because of the higher angle of attack. Meanwhile, the downward force on the tail will be reduced because of a decrease in its negative angle of attack. The increase in lift on the wing and the reduced downward force on the horizontal stabilizer result in the center of lift moving aft. The airplane is no longer balanced but is stable because there is a restoring torque. Thus the airplane will rotate back toward the balanced condition, where it will resume straight-and-level flight.

The toy airplanes in the center of Figure 3.2 illustrate what happens when the center of gravity is exactly at the neutral point. Notice that the center of gravity is slightly aft of the center of lift of the wing alone. Thus, to balance, the horizontal stabilizer must produce a small positive lift. In this particular situation, the center of lift is independent of the angle of attack. In other words, when the airplane is perturbed from straight-and-level flight, the increased lift on the wing and the increased lift on the tail balance such that the center of lift remains stationary. This happens at the unique location where the wing's lift and tail's lift change together in exact balance. Thus, when the airplane pitches up owing to some disturbance, as shown with the toy airplane on the right, there is no restoring torque about the center of gravity. This toy airplane is neutrally stable and balanced.

Finally, the third example in Figure 3.2 shows what happens if the center of gravity is behind the neutral point. In straight-and-level flight, as shown on the left, the lift on the wing is balanced by the lift on the tail. However, in its perturbed state, for example, 5 degrees nose pitch up, the lift on the wing grows faster than the lift on the tail. The result is a rotational torque that rotates the toy airplane farther from its initial straight-and-level flight state. This is, of course, unstable.

> A KLM DC-2, which was making a commercial passenger flight, came in second in the 11,333-mile race from England to Australia in 1934.

One question that you might ask is how the lift on the wing can grow faster than that on the tail because lift is just proportional to the effective angle of attack. There are two reasons. First, the tail is generally less efficient. On a typical airplane, the tail has about half the aspect ratio of the wing. The lift grows less rapidly with angle of attack for smaller-aspect-ratio wings. The other factor that affects the horizontal stabilizer's lift is that it is flying in the downwash of the wing. If the wing is generating more lift, it has a greater downwash. The net effect of the downwash is that the horizontal stabilizer sees a lower relative angle of attack than it would if there were no downwash. The wing may experience a 10-degree change in the angle of attack, whereas the horizontal stabilizer only sees a 6-degree change in the direction of the air.

The lesson is that for longitudinal (pitch) stability, it is crucial that the center of gravity of an airplane be forward of the neutral point of the airplane. That is, the plane is nose-heavy. Such an airplane, when disturbed by a gust of air or a sudden control movement, will tend to return to the original attitude. If the center of gravity is behind the neutral point, the airplane is not flyable. Any disturbance will be magnified and will tend to increase. Thus, if the nose pitches up just a little, the airplane will want to exaggerate this motion. In such a situation, the controls will not respond at all. Therefore, the airplane must be loaded ahead of the neutral point.

Pilots must determine the center of gravity before each flight, depending on fuel, passengers, and payload, to ensure that the airplane's center of gravity is within limits. Unfortunate accidents have resulted from pilots inadvertently taking off with an airplane loaded such that the center of gravity was behind the neutral point. In this situation, no effort by the pilot can save the airplane from catastrophic consequences.

There is another reason to have the airplane's center of gravity forward of the neutral point, and that is for stall recovery. When a stall occurs, if the center of gravity is too far aft, the airplane will be tail-heavy, and the angle of attack will increase after the stall. The airplane then would flip. When the airplane is nose-heavy, the nose will drop automatically in a stall, reducing the angle of attack, and thus recover from a stall.

We have discussed only the horizontal stabilizer with respect to stability. One must remember that the elevators are used for pitch control but also to adjust the lift of the horizontal stabilizer for various loadings of the airplane.

Trim

An airplane must be able to be balanced and stable for a variety of loads. For example, a commercial airplane may face two extremes, one where the airplane is half full of passengers who are all sitting in the forward seats and one where the airplane is half full and all the passengers are sitting in the aft seats. The two seating arrangements will move the center of gravity of the airplane. We assume that in both cases the center of gravity is ahead of the neutral point so that the airplane is stable. However, the airplane also must be balanced. This is where elevator trim plays a role.

Depending on where the center of gravity is, the horizontal stabilizer's lift must be adjusted to balance the airplane. The pilot could achieve this by holding the elevator in position, but this would be tiring on a long flight. Instead, therefore, the pilot trims the airplane to balance it. In many airplanes, a small flap on the elevator, called a *trim tab*, is adjusted from the cockpit so that the horizontal stabilizer produces the right amount of force to balance the airplane without moving the elevator, as shown in Figure A.2 in Appendix A. On many airplanes, such as the large commercial transports, the entire horizontal stabilizer is rotated so that its angle of attack can be adjusted for different flight conditions.

> For a typical international flight, one Boeing 747 operator uses no fewer than 5.5 tons of food supplies and more than 50,000 in-flight service items.

Now, what happens when passengers move about in the airplane? The center of gravity moves, and the airplane will want to pitch nose up or down. Thus the horizontal stabilizer trim is adjusted to compensate. However, if the pilot wishes to purposely change the pitch, or angle of attack, he or she will command the elevators. Thus the horizontal stabilizer and trim tabs maintain pitch stability, and it is the elevator that controls the pitch.

Flying Wings

Today, there are examples of airplanes that have no horizontal stabilizers, such as the B-2 bomber shown in Figure 3.3. How can this be? It would seem impossible for the airplane to compensate for any shift in loading. If the B-2 copilot decides to move to the aft cabin, will the B-2 flip? This is the tricky part of designing a flying wing. In this case, control surfaces on the trailing edge of the wing move up or down in unison, like an elevator, which moves the center of lift forward or aft depending on the location of the center of gravity. These control surfaces both balance and stabilize the airplane.

Figure 3.4 shows a Northrop YB-49 Flying Wing, which flew in the late 1940s. In the figure you can see the two control surfaces, both in the down position, allowing them to act as elevators and flaps. They

FIGURE 3.3   B-2 bomber. (Photograph courtesy of the U.S. Air Force.)

FIGURE 3.4   Northrop YB-49 Flying Wing.

are also able to act in an opposite direction of each other like ailerons to give roll control.

Horizontal Stabilizer Sizing

How big should the horizontal stabilizer be? There are two primary ways to increase the effectiveness of a horizontal stabilizer. One is to increase the stabilizer's area, and the other is to increase its distance from the wing. In fact, it is the distance to the horizontal stabilizer times the wing area of the horizontal stabilizer that dictates stability. Two stabilizer configurations, one with half the area but twice as far from the wing, will have approximately the same stabilizing effect. Note that the effect of downwash also must be considered because the stabilizer that is twice the distance will experience less of the wing's downwash.

The neutral point of an airplane is a function of the horizontal stabilizer lever arm times the area. The greater this product is, the farther aft is the neutral point. The advantage to having a neutral point farther aft is that it gives an increased range on center-of-gravity location. This gives the airplane greater loading flexibility.

On the other hand, there is a negative impact of increasing either the lever arm or the area of the horizontal stabilizer. Both translate into more weight, the airplane designer's nemesis. Both also contribute to higher drag owing to skin friction. Another problem can be that the

> Many people did not believe that the Wright brothers had flown in 1903. Until their demonstration in France in 1908, the French called them the "Wright liars."

airplane becomes too stable. In other words, the effectiveness of the horizontal stabilizer is so great that the pilot has a difficult time changing the pitch angle. This comes under the category of handling properties, which will be discussed later. Therefore, sizing a horizontal stabilizer involves a combination of versatility in center of gravity, weight, drag, and handling properties.

Directional Stability

In the preceding section we discussed only stability in pitch, known as *longitudinal stability*. Appendix A introduces you to two other axes, roll and yaw. Roll stability, known as *lateral stability*, was covered in detail in Chapter 2. The effects of dihedral and sweep were presented as well and will not be repeated here. *Directional stability* is the stability in the yaw axis and gives rise to the vertical stabilizer. The vertical stabilizer and rudder serve the same function as the horizontal stabilizer and elevator, except in yaw instead of pitch. The main function of the vertical stabilizer is to help the airplane *weathervane* and thus keep the nose pointed into the direction of flight.

The desire for directional stability is to have the airplane always line itself with the oncoming air. Thus, if a gust temporarily perturbs the direction the nose is pointed, the tail will have a nonzero angle of attack with the airflow, as shown in Figure 3.5. This causes a restoring force, actually horizontal lift, to realign the tail with the direction of travel. The effects of misalignment with the flight path are primarily high drag and poor turn coordination.

The size of the vertical stabilizer depends on several factors. For a single-engine airplane, the requirement that sets the minimum size for the vertical stabilizer is that the torque from the vertical area of the airplane aft of the center of gravity be larger than the torque from the vertical area forward of the center of gravity. This is the same requirement that puts feathers on arrows for stability. A larger vertical stabilizer is needed to counter propeller rotation effects and adverse yaw in a turn. A single-engine airplane can get by with the minimum size vertical stabilizer, but it will require more work on the pilot's part.

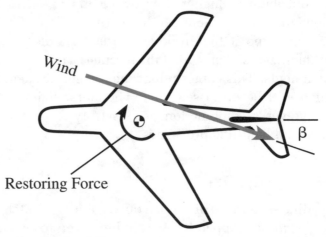

FIGURE 3.5 Directional stability.

For multiengine airplanes, the size of the tail is dictated by the torque caused by the loss of one engine. The net thrust being off center causes the airplane to want to yaw. A large vertical stabilizer, with trim, can compensate for this. This is why twin-engine commercial transports have such large vertical stabilizers.

The Federal Aviation Administration (FAA) dictates limits on directional stability. Modern airplanes now have vertical stabilizers that are so effective that they make use of the rudder for small corrections almost unnecessary.

> To remain stable, the flight control system on the X-29 (see Figure 2.8) has to make 40 corrections per second.

The B-2 bomber, which has no vertical stabilizer, accomplishes directional stability by using wingtip drag. The ailerons on the side to which you wish to turn will split, deploying both up and down. Thus they increase the drag on that side, pulling the nose around. The ailerons are seen split on both wings in Figure 3.3.

Dynamic Stability

Static stability deals with the tendency of an airplane, when disturbed, to return to its original flight attitude. Dynamic stability deals with how the motion caused by a disturbance changes with time. The three types of dynamic stability are shown for the longitudinal stability of an airplane in Figure 3.6. The first flight path shows positive dynamic stability. When the airplane pitched up, there was a restoring force (statically stable). The path oscillates through the original altitude and with the oscillations decreasing with time. This is like a car with good shock absorbers.

The second case is neutral dynamic stability. The airplane is statically stable because there is a restoring force. But the amplitude of the oscillations in this case does not decrease with time. This is kind of like a car without shock absorbers that hits a bump.

The third case in the figure is negative dynamic stability. Again, the airplane is statically stable, but the amplitude of the oscillations increases with time. This is kind of like a car without shock absorbers going down a "washboard" (bumpy) road.

As with static stability, dynamic stability must be considered in the longitudinal, lateral, and directional (pitch, roll, and yaw) axes. What

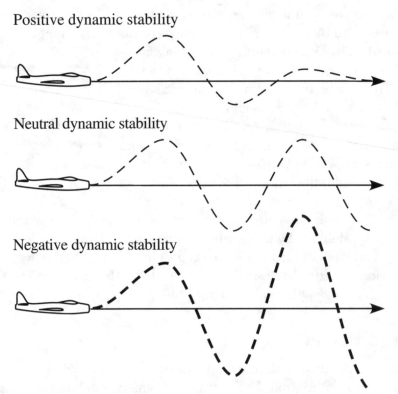

Positive dynamic stability

Neutral dynamic stability

Negative dynamic stability

FIGURE 3.6   Three types of dynamic stability.

can make dynamic stability even more interesting is that more than one axis can couple, producing some very interesting motion. In the following sections you will be introduced to three modes of dynamic motion that are the easiest to understand and may be most familiar to you.

Phugoid Motion

Have you ever thrown a paper airplane and watched it follow a flight path that climbs and slows and then descends and speeds up, as illustrated in Figure 3.7? This type of motion is common to all aircraft and is given the name *phugoid motion*. Phugoid motion is a trade between kinetic and potential energy, that is, speed and altitude. It occurs at a constant angle of attack so that as the speed increases, so does the lift. The extra lift causes the airplane to increase altitude. As it does, the airspeed falls off, decreasing lift and, eventually, altitude. In the extreme, at its maximum height, the airplane will lose so much speed that it will stall.

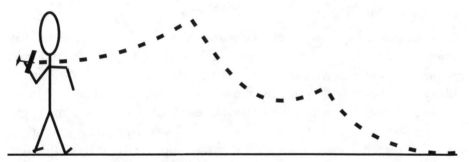

FIGURE 3.7   Phugoid motion.

There is no need to worry about this in a real airplane. The time it takes to complete one cycle, or period, is on the order of minutes. Only a sleeping pilot would ignore this. In fact, the period is so long that most pilots do not even recognize that they are controlling this motion. Also, the oscillations will damp out eventually by themselves if they are ignored.

Dutch Roll

The *Dutch roll* is a motion that couples roll and yaw. The name comes from the motion of the Dutch speed skaters as they glide across the ice. It is kind of like the coupling of a small rolling motion with a small wiggle of the tail. The airplane does this while maintaining its heading. The Dutch roll has a short period and presents no major stability problem. The biggest problem is upset stomachs of passengers in the rear of the airplane. A larger vertical stabilizer will help dampen Dutch roll. Because of passenger discomfort, commercial airplanes use *yaw dampers* that move the rudder to automatically dampen any Dutch roll. A yaw damper is not a critical flight control, so a flight will continue if it is inoperable. There may be more sick passengers on the flight, though.

> During the Gulf War, U.S. airplane loss rates were roughly the same as during normal training.

In smaller airplanes, most pilots do not feel the motion of a Dutch roll. This is so because all motion occurs around the center of gravity. In a small airplane, the pilot and passengers are usually sitting very close to the center of gravity. Only on a large commercial transport do passengers find themselves far enough away from the center of gravity to feel this motion.

Spiral Instability

The last instability that we will consider is in spiral motion, frequently referred to as *spiral divergence*. This is instability in yaw and roll, which leads to a downward spiral. An airplane that has a spiral instability eventually will spin into the ground if the pilot does not intervene. Although the results are dramatic, the FAA allows for some spiral divergence in airplane designs. This is so because, like phugoid motion, spiral divergence takes a long time to develop. Again, only a sleeping pilot could ignore this. Note that Lindbergh's unplanned nap might have been disastrous if he had not woken up on time. The *Spirit of St. Louis* was fairly deep into a downward spiral at the time he woke up.

Stability Augmentation

The newest military aircraft designs are exploring statically unstable airplanes, which could not be flown without computer control. What the computer does is kind of like when you try to balance a pencil vertically on the tip of your finger. It is very difficult. This is a statically unstable situation. However, if you could react quickly enough, you could keep the pencil in position. Quick reaction is the role of the computer in the unstable airplane.

Why would the military want a statically unstable airplane? As discussed in Chapter 2, the answer is maneuverability. If the airplane has a natural tendency to diverge from a specific condition, such as straight-and-level flight, then it will be much more responsive when the pilot wants to make a change. Another reason for designing a statically unstable airplane is that smaller stabilizers might be used, which decreases the weight and drag of the airplane.

> Northrop designed a high-speed flying wing, the XP-79, that was a flying ram. Its objective was to slice off the tail of the opponent with its leading edge.

The ability of a computer to solve problems quickly gives it a tremendous advantage. The Wright brothers preferred their airplanes slightly unstable so that the pilot would have to react and pay attention. But the design was only slightly unstable, and the airplane was controllable by use of the control surfaces. If an airplane were highly unstable, the pilot would not be able to react fast enough to compensate. Today, inserting a computer into the control loop can augment stability. The pilot can manage the overall flight path while the computer manages the quick-response tasks, or the computer can do both.

With the computer in the control loop, an airplane can be built to be naturally unstable. One of the first examples of an unstable design was the X-29, shown in Figure 2.8. The computer makes fast, tiny adjustments that allow the pilot to focus on other tasks. If the computer were to fail, the airplane instantly would become uncontrollable to the pilot, with disastrous consequences.

Handling

As mentioned earlier, handling qualities were not quantified until after World War II. Before that time, handling qualities relied on pilot opinion. Words such as *hot*, *fun*, *smooth*, *fast*, *sluggish*, and *sporty* are still used by pilots to qualify an airplane's handling properties. But what is *sporty* to one pilot might be *sluggish* to another. The more dangerous situation is the opposite, where the high-time fighter pilot, who considers the airplane to be *smooth*, turns the airplane over to a novice, who finds it rather *sporty*. This is a particular problem in the home-built industry, which rarely publishes quantitative handling data.

One handling quality is *stick force*. This is a measure of how much force is required to make a certain change in a control surface. Suppose that an airplane required 40 lb of force to roll the airplane at 1 degree per second. This would qualify as *extremely sluggish*. However, if a 1-lb force on the controls corresponds to a roll rate of 180 degrees per second, this would be *very sporty*.

Another issue is *control balance*. Suppose that you have to put 5 lb on the control yoke for maximum roll but 30 lb for maximum pitch. This is an unbalanced control system. Ideally, 5 lb on the control yoke should give roughly the same changes in both the roll and the pitch axes.

Another factor is the *adverse yaw*. Older airplanes had significant adverse yaw, so a pilot had to be diligent with rudder pedals. A modern trainer hardly needs any rudder input to counter adverse yaw. The

> On modern commercial transports, the flight management computer can control the overall flight path. In fact, a pilot can preprogram the flight from takeoff to landing and never have to touch the controls. The computer can fly with such accuracy that airplane wheels will land at the same point on the runway, which eventually can wear out that spot. Dispersion is introduced into the autoland feature so that touchdowns will occur over a broader area.

improvement has come primarily through the use of dihedral and larger vertical stabilizers. Older pilots consider this sloppy flying.

Fly by Wire

Before the days of the digital computer, airplane control surfaces were linked to the control yoke through cables, push rods, and hydraulic lines. These were mechanical links from the pilot's controls to the control surface. In the case of cables and push rods, the stick force was a matter of designed mechanical advantage that was using the basic concept of a lever. The problem with cables and push rods is that they can be difficult to route from the yoke to the control surface. For example, a cable or push rod would be undesirable down the center of the cabin. Hydraulic lines, on the other hand, can be routed fairly easily because they are just tubing.

With the computer in the loop, there is no need for direct mechanical connections or running hydraulic lines through the airplane. A fly-by-wire system is a control system where control actions are transmitted by wire. The pilot inputs a command on the yoke, which is read by a computer. The computer translates the command, along with its own inputs to augment stability, to an electrical signal. A wire then connects the cockpit to various *actuators*, which convert the signal into a mechanical action, such as moving the elevator.

> DC-6 publicity photos used a model named Norma Jean Baker inside the cabin. She later became known as Marilyn Monroe.

There are a few interesting side effects of fly-by-wire systems. One is that an intelligent computer can be used to make decisions. For example, the computer may monitor the angle of attack and not allow the aircraft to reach the stall angle of attack. Thus, no matter how hard the pilot pulls back on the control yoke, the airplane will not increase its angle of attack to a stall. This can be useful in a fighter aircraft in which the pilot in combat may be preoccupied with the fight. Another example is that the computer might turn a sloppy pilot-controlled landing into a smooth landing. In essence, the pilot and the computer both fly the airplane.

Another side effect worth noting is the effect on stick force. Flying an airplane with a joystick is no different from flying in a computer simulation with a joystick. The joystick on the computer has no way of giving mechanical feedback in terms of resistance to turning. Thus it takes the same force to make a tight turn at low speed as at high speed. This feedback is a point of contention between pilots and designers. The pilots want the mechanical feedback. Now, at least one airplane manufacturer has developed springs and linkages that are controlled by the computer system and give the pilots the same feel as the old mechanical linkages. Of course, now this means

> Howard Hugh's infamous *Spruce Goose* was actually made mostly of birch.

that the stick force and balance can be tuned with software. At the same time, though, this is extra hardware, and thus weight, that only serves the purpose of making the pilot more comfortable. However, safety in itself provides a justifiable argument in favor of such systems.

The computer has resulted in an interesting aspect of handling properties. With the computer installed between the pilot and the control surfaces, the properties of stick force and balance can be changed in software. This can be useful in a new-airplane development program. For example, the expected handling properties of the Boeing 777 were programmed into a Boeing 757, which was flown extensively by the test pilots before the first Boeing 777 was even built. The first flight of the Boeing 777 went off without a hitch in part because the pilot had many simulated hours flying a Boeing 777.

Wrapping It Up

In this and the preceding chapter you were introduced to basic concepts of stability and control. There is a great deal more detail that could be covered on this topic, but you have been given a basic introduction as a starting point. In the early days, stability and control were not as significant a part of the airplane's design as they are today. On a modern airplane, both commercial and military, the stability and control systems include computers. Computers have many capabilities, and engineers are learning more ways to use them. Thus, in addition to augmenting stability, an airplane's computer may take navigation data, instrument landing data, air traffic data, and weather data and create and execute a flight path. In fact, the computer alone can fly an entire flight, from takeoff to landing. The pilot's job becomes more of a systems monitor than the person flying an airplane. There is a joke that a pilot and a dog will fly the airplanes of tomorrow. The dog's job will be to bite the pilot's hand if he touches the controls, and the pilot's job will be to feed the dog.

Airplane Propulsion

The propulsion system is one of the most complex systems on an airplane, yet the principles behind airplane propulsion are not very complicated. An airplane in flight requires power to provide lift and to overcome the drag associated with the impact of the air on the airplane. If it is climbing or making a turn, additional power is needed. In Chapter 1 we showed that in producing lift, work is done on the surrounding air. Here we will use similar arguments to describe how aircraft propulsion systems work.

The power required for climbing and turning will be discussed in Chapter 6, whereas in this chapter we will see how propulsion systems provide the necessary power for flight. We will also show how jet engines differ from piston engines in how they deliver the needed power. Some of the differences and similarities may be surprising.

With some exceptions, gliders being the most notable, the power needed for flight is provided by either a piston or a jet engine. The energy produced in these engines must be transferred to the surrounding environment to propel the aircraft. These propulsion systems require some very complex engineering. However, one need not understand the details of how those systems work to understand what the propulsion system is doing and why engines look the way they do.

It's Newton's Laws Again

We saw in Chapter 1 that a wing creates lift by diverting air down. The same principles explain how an aircraft propulsion system works, except that to create thrust, it must push air back. Like a household fan that pushes air, a propeller and a jet engine do the same. More important, like wings, aircraft propulsion systems are applications of Newton's laws.

> The world's longest recorded flight of a chicken was 13 seconds. The greatest known distance a chicken has ever flown is 301.5 ft (92 m).

Remember that Newton's third law states that *for every action, there is an equal and opposite reaction*. In an aircraft propulsion system, the action is the acceleration of air or exhaust, and the reaction is the force, or thrust, produced. Again, we will use the alternate form of Newton's second law, which states that *the thrust is proportional to the amount of gas accelerated per time times the speed of that air*.

Thrust

Although not a conventional aircraft propulsion system, a rocket motor is a good starting point for understanding propulsion. An example of how a rocket motor works is shown in Figure 4.1. Fuel and an

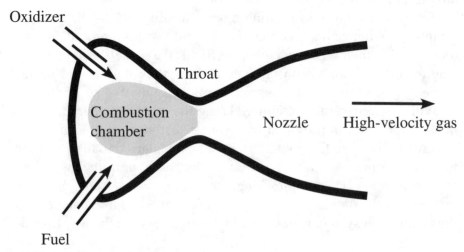

FIGURE 4.1   Schematic of a rocket motor.

FIGURE 4.2   Apollo 8 launch. (Photograph courtesy of NASA.)

oxidizer are pumped into a combustion chamber, producing a large amount of gas at high pressure. The gas accelerates to the throat of the motor, where it reaches a velocity of Mach 1; in other words, equal to the speed of sound. After the throat, the gas continues to accelerate, producing an exhaust of hypervelocity gas with a great deal of thrust. A large rocket expels a great deal of gas at high velocity. The magnitude of the gas expelled can be seen in Figure 4.2, which shows the launch of Apollo 8.

The thrust from a rocket motor is analogous to the recoil of a rifle when a bullet is fired. From the alternative statement of Newton's second law, we say that the thrust of a rocket motor is proportional to the amount of mass expelled per time times the velocity of the gas. In order to increase the thrust of a rocket, one can increase the amount of gas expelled per second, the velocity of that gas, or both. An aircraft propulsion system works much the same way as a rocket motor. The force on the airplane is a reaction to accelerating air or exhaust. If you have ever stood behind an airplane with its propeller turning, you certainly can recognize that a great deal of air is being blown back.

Power

Aircraft propulsion consists of two distinct parts. There is the engine that converts a source of energy, such as fuel, to work. Then there is the part of the system that converts the work of the engine into work on the surrounding environment to produce thrust. A piston engine plus propeller combination is an example of a complete aircraft propulsion system. A turbojet is another example, but the parts of that system are a little harder to distinguish from each other.

Usually, engineers, flight instructors, and educators relate flight and propulsion in terms of forces. In this book, we take the perspective of power, which is adjusted by the throttle, can be measured by the pilot, and is more intuitive. If one increases the throttle or fuel flow, the power increases accordingly. Power is the rate of using energy, or doing work, which is the key to understanding propulsion. Power also lends itself to another fundamental concept: efficiency.

Looking at the propulsion system this way, it is convenient to introduce a few terms. You know that power is required for flight: for supporting the weight of the airplane, for overcoming friction, for climbing, etc. This is the *power required* for flight. The power that is actually produced by the engine and delivered to the propeller or is available for propulsion by the jet we will call the *power available*. The difference between the power available and the power delivered by the engine we will call *wasted power*. This is primarily the power that is lost to kinetic energy in the propeller's slipstream or the jet's exhaust.

On May 9, 1926, Commander R. E. Byrd made the first flight over the North Pole. On November 28–29, 1929, Byrd made the first flight over the South Pole.

The power available is the thrust times the speed. With a piston engine, for a fixed engine power, the power available depends only slightly on the airplane's speed. Figure 4.3a illustrates how the thrust and power available vary with speed for a typical piston engine plus propeller combination at a fixed power setting. The thrust from a propeller degrades with speed, but the power available

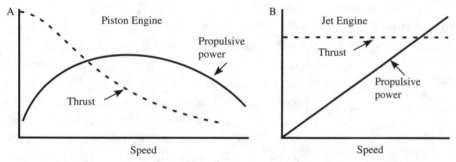

FIGURE 4.3   Power available and thrust as functions of speed for a propeller and a jet engine.

holds up quite well. Of course, the details depend on several factors, propeller design being one of them.

As can be seen in Figure 4.3b, the jet engine behaves quite differently with speed than a piston engine. The thrust available to a jet engine is roughly constant with speed, and the power available is proportional to the speed of the airplane. This will be discussed in more detail later, although one can anticipate that the differences will affect the performance of an airplane.

Efficiency

The objective of an aircraft propulsion system is to produce the required power as efficiently as possible. There are basically two areas where propulsion systems lose efficiency. The first is in the conversion of fuel to engine power. This is the *engine thermal efficiency*. The losses here are primarily due to inefficiencies in the burning of the fuel and energy lost to heat and friction within the engine. Energy is also used for engine-system support, such as powering fuel and lubricant pumps and generating electricity. These losses reduce the engine's output efficiency.

Once the engine has converted the chemical energy of the fuel into mechanical energy, it must convert the mechanical energy into propulsion. The ratio of the power available to the engine power gives us the *propulsive efficiency*. The total efficiency of the propulsion system is a measure of how much power the system develops for a certain quantity of fuel burned. So what contributes to the propulsive efficiency? The same arguments used for lift efficiency can be used here.

Remember from Chapter 1 that the lift of a wing is proportional to the momentum (mv) that is transferred to the air per time. The kinetic energy left behind in the air by the passing of the airplane ($\frac{1}{2}mv^2$) is an energy loss to the airplane. For the most efficient flight, one wants to produce the necessary lift while leaving as little energy in the air as possible. Therefore, one wants to divert as much air as possible at as low a velocity as possible. This is why the efficiency of a wing increases with size. Increased lift efficiency requires an increase in the amount of air diverted, not an increase in the speed of the diverted air.

> Before their first flight in a powered airplane, the Wright brothers had accumulated more flight time in gliders than had been accumulated in history before them.

Propulsion systems produce thrust and deliver their power to the surrounding environment in the same manner as a wing. Thus, for the most efficient thrust, one wants the engine to accelerate as much air as possible at as low a velocity as possible. This minimizes the wasted power. If a propeller or a jet could discharge a very large amount of air or exhaust at a relatively low velocity, it would take much less power than another system that developed the same thrust by discharging a small amount of air at a high velocity. Keep in mind that the energy given to the air producing the thrust is lost energy that must be paid for by the engine.

For the best propulsive efficiency, one would want to have almost zero kinetic energy left behind in the air. Unfortunately, this is not possible. To produce thrust, there always must be some velocity given to the air and thus some waste.

Let us look at the situation of an engineer who wishes to design a jet engine that is to have twice the thrust of a previous engine. The engineer can increase the thrust by increasing the amount of mass discharged by the engine per time, by increasing the velocity of the exhaust, or by increasing both. However, if one increases the exhaust velocity, the wasted power increases as the velocity squared. So doubling the thrust by doubling the exhaust velocity doubles the power available but increases the wasted power by a factor of 4! However, if one doubles the thrust by doubling the mass discharged, the power available again has been doubled, and the wasted power has only doubled. Basically, the objective of an aircraft propulsion system is to create the most thrust for the least wasted power, which leads aircraft engine designers to favor increasing mass flow over exhaust velocity.

Even for the most efficient propulsion systems, a great deal of kinetic energy is imparted to the air. There is no ideal propulsion system for flight. Wings are much more efficient at producing lift than engines are at producing thrust. This is so because wings, with their large size, are able to divert a great deal of air at a slow speed. Engines do not have this luxury.

Propellers

A propeller is simply a rotating wing. For low-speed flight, a propeller is the most efficient means of propulsion. At peak efficiency, 84 percent or more of the engine power can be converted into power available by a propeller. Thus only about 16 percent of the power is wasted.

Since the propeller works by pushing a certain amount of air mass through each revolution and accelerates this mass of air to a higher velocity, we can begin to understand the tradeoffs in propeller design. A large-area propeller will be more efficient than a smaller one because it can push more air. Rubber band–powered model airplane enthusiasts know that a large, slow-turning propeller will result in longer flights.

Propeller size and rotation speeds are dictated by many factors. First, a large, slow-turning propeller may not be practical because of ground clearance. Of equal importance, though, is matching the propeller rotation speed to the type of engine provided. Couple these requirements with the need to keep the propeller-tip speed below the speed of sound (for reasons of noise and additional power loss), and you have the propellers we see on airplanes today. Airplane piston engines usually are designed to operate at between 2200 and 2600 rpm (revolutions per minute). Limiting the tip speed to roughly the speed of sound in normal flight situations gives a typical propeller diameter of 72 to 76 in (182 to 193 cm). Very early airplanes, such as those used in World War I, had slower-turning engines, operating at about 1600 rpm. These early airplanes had very large propellers (100 in [255 cm]). Because the larger diameter allowed them to push more air per second, they had fairly high propulsive efficiency, comparable to what we can do today. However, engine efficiency has come a long way.

The power added to the air is proportional to the propeller's rotational speed cubed. This can be easily understood by considering a wing with a fixed angle of attack. We know from Chapter 1 that the power imparted to the air by the wing is proportional to the amount of air diverted times the vertical velocity of that air squared. If the speed of the wing is doubled while keeping the angle of attack constant, both the amount of air diverted and the vertical velocity of that air also will be doubled. Thus the power will have gone up by a factor of 8. Likewise with a propeller, with a fixed pitch and constant forward speed, the power transferred to the air is proportional to the propeller's rotational speed cubed. What this means is that the power needed to turn a propeller increases very rapidly with its rotation speed. It is therefore very important to get the area of the propeller correct for the engine size. If the blade is too small, the load on the engine will be low, and the engine will "over-rev" by going to a high rpm value. This can damage an engine. If the blade is too large, the engine will not be able to reach its optimal operating speed and thus will not be able to deliver full power to the propeller.

> The rocket for the Apollo 4 launch was the noisiest of all rockets. It was so loud that it produced seismic readings in New Jersey, a distance of 850 mi (1400 km).

Multibladed Propellers

The total surface area of the blades on a propeller determines the ability of the propeller to convert the engine's power into thrust. The more area, the more power the propeller can convert. For most small planes, the appropriate blade area is achieved with two blades, with the area of the blades becoming greater with increasing engine size. Some propellers increase the total area by using three-, four-, and even six-bladed propellers. The reality is that if the total area is the same for two-bladed versus three-, four-, five-, or six-bladed propellers, the efficiency will be close to the same.

Transitioning from two to more blades with the same total area is a result of subtle tradeoffs. Two-bladed propellers are usually best for lower-speed airplanes, where the power requirements are low. More blades are used when power requirements are higher, such as for faster climbing and higher speeds. Other factors favoring multibladed propellers are that they produce less objectionable noise and reduced vibrations. So why not always use more blades?

FIGURE 4.4 One-bladed propeller. (Courtesy of Pearson Air Museum, Vancouver, WA.)

One reason for not using more blades is that they are more expensive. Another reason is that for more blades, there is more disturbed air that an individual blade sees, and thus there is a less efficient transfer of power. Thus, in general, the fewer blades that one can use, the better. This has been taken to the extreme by efforts to use one-bladed propellers, as shown in Figure 4.4. It has been shown experimentally that the turbulence in the slipstream of a blade is greatly reduced in one revolution. This improves the propeller's efficiency. However, since propellers are already quite efficient, the improvement is not of great interest.

Propeller Pitch

With the size and speed of the propeller fixed, the airflow through the propeller is basically fixed. Therefore, in order to get more thrust, the air velocity behind the propeller must be increased. The pitch of a propeller is analogous to the wing's angle of attack. With a *fixed-pitch* propeller, the angle of the blades is fixed with respect to the propeller's rotation direction. The propeller's apparent angle of attack is determined by the *pitch* of the propeller, its speed of rotation, and the speed

In Figure 4.5, the direction of the apparent wind and the direction of lift on the propeller are shown. All angles are exaggerated to make the figure easier to understand. Remember that the lift on a wing or a propeller is perpendicular to the direction of the oncoming air as seen by the airfoil. With no forward speed, the lift on the propeller is forward. As the airplane increases its forward speed, the lift tilts so that not all the lift force is in the direction of travel. This creates an inefficiency that increases with forward speed.

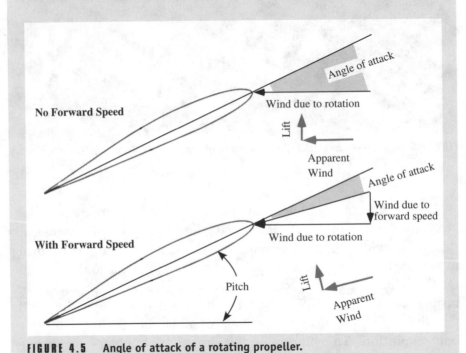

FIGURE 4.5   Angle of attack of a rotating propeller.

of the airplane through the air. The faster the airplane is traveling, the smaller is the apparent angle of attack. This is illustrated in Figure 4.5, which shows the angle of attack for a propeller on an airplane with no forward speed at run-up and in flight. As the airplane flies faster, the wind, owing to the airplane's forward speed, reduces the angle of attack of the propeller. Thus the propeller diverts less air, producing less thrust and requiring less power from the engine.

The efficiency of a fixed-pitch propeller depends on the speed of rotation and the speed of the airplane. Figure 4.6 shows the efficiency

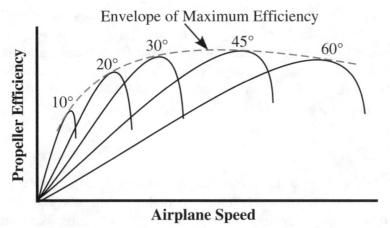

FIGURE 4.6 Efficiency of a propeller as a function of the speed of the airplane.

of a propeller at a single rotation speed for various pitch angles as a function of the speed of the airplane. From this, it is clear that for a single pitch, the efficiency is optimal over a fairly narrow range of airplane speeds. Because of this, a fixed-pitch propeller must have a fairly high pitch for all-around performance. The rated engine power available at a given altitude is determined by the engine's rpm value, which is also the propeller's rotation speed. Thus high pitch may cause the engine to run below its optimal speed during takeoff and not produce full power. The same blade at cruise speed may require that the engine be throttled back to prevent the engine from operating at too high of an rpm value. Thus the fixed-pitch propeller is a compromise.

A solution to this compromise is the *constant-speed propeller*. The constant-speed propeller allows the pilot to control both rotation speed and propeller pitch. A constant-speed propeller works something like a governor on an engine. There are two controls, the engine throttle and an rpm control. The throttle controls the power output of the engine, and the rpm control sets the rotation speed of the propeller and thus the speed of the engine. If the engine wants to run too fast, the pitch of the propeller is automatically increased until the engine slows down to the preset speed. This allows the efficiency of the constant-speed propeller to look something like the *envelope of maximum efficiency* shown in Figure 4.6.

On takeoff and in climb, the propeller is adjusted to have a fairly small pitch. Because of the airplane's slow speed, the angle of attack is still quite large. In cruise, the airplane's speed causes the propeller to see a reduced angle of attack. Here, the pitch of the propeller is increased to allow the engine to operate at its optimal performance. Typical constant-speed propellers on small airplanes improve the fuel efficiency at cruise by about 14 percent as well as improving the power available on takeoff and climb.

Piston Engines

Either a piston or a turbine engine can power a propeller. Here, you will be briefly introduced to aircraft piston engines. We will not go into detail about how they work but rather focus on the relationship of the piston engine to the power delivered.

In brief, a piston engine works by converting a certain amount of energy from a fuel and oxygen mixture into kinetic energy of a piston. The energy of the piston then is used to turn a shaft. Finally, the shaft turns a propeller. This is illustrated in Figure 4.7.

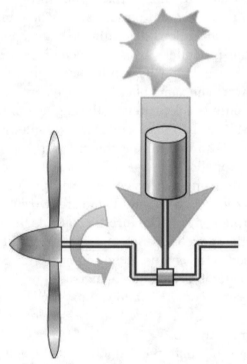

FIGURE 4.7 Piston engine converting chemical energy into propeller rotation.

If we define the *total power* as the amount of energy per second available in the fuel and air mixture, then the total power will be limited by the amount of air and fuel that can be pumped into the cylinders of a piston engine. Larger cylinders mean a more powerful engine. The higher the altitude, and thus the lower the air density, the less power is produced. Because the available oxygen decreases with altitude, the *normally aspirated* piston engine generally limits aircraft to low operating altitudes.

One way to overcome the power loss with altitude is to add a pump to the air intake to increase the amount of air in the cylinders. There are two common methods for doing this. The first is called *turbocharging*. A turbocharger makes use of the energy expelled in the exhaust to run a small pump in the air intake. A *supercharger* is another method to pump additional air into the cylinders. A supercharger can be powered mechanically through a belt on the engine shaft or with an electric motor. The purpose of both is the same, to increase the amount of air (oxygen) in the cylinders at higher altitudes where the air is less dense. The result is that a turbocharged or supercharged engine can maintain constant power up to a higher altitude. Above that altitude, the pump can no longer main-

> In Sioux City, Iowa, on July 19, 1989, a DC-10 landed after losing all tail surface controls owing to an explosion in the middle tail engine. The pilot miraculously maneuvered the airplane to a crash landing using the thrust from the two wing engines to turn.

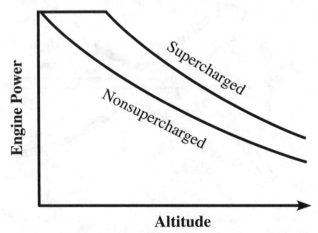

FIGURE 4.8   Engine power as a function of altitude for a nonsupercharged and a supercharged engine.

tain sea-level density in the cylinders, and the power drops off with further increase in altitude. Figure 4.8 shows engine power as a function of altitude for a nonsupercharged engine and for a supercharged engine. Turbochargers or superchargers usually are not used to increase engine power at sea level. Engine temperatures and loads would be too great, resulting in damage to the engine.

The thing to keep in mind with piston-engine-powered airplanes is that the engine's power is a function of altitude but not the speed of the airplane. As shown in Figure 4.3, thrust of the propeller decreases with speed, but the power available is pretty much constant until high speeds are reached.

Turbine Engines

A turbine engine, which is the heart of all jet engines, is very different from a piston engine. For one thing, it is a little harder to separate the "engine" from the device that produces thrust. We will start this discussion by introducing the basic elements of the turbine engine in what is called the *engine core*.

In a turbine engine, the energy of the combustion is transferred to the exhaust rather than to a mechanical piston. Figure 4.9 shows how

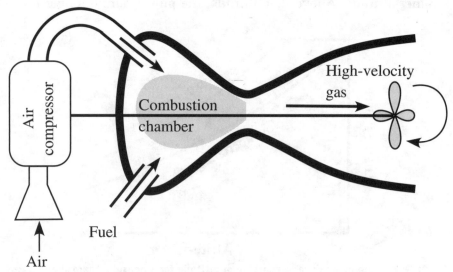

FIGURE 4.9   Concept of how a turbine engine works.

one might conceptualize a turbine engine, starting from the rocket motor in Figure 4.1. The rocket carries fuel and its own source of oxygen. This is how it is able to operate in the vacuum of space. But a turbine engine flies in the atmosphere where oxygen is plentiful. So the rocket motor could be fitted with a compressor to supply high-pressure air to the combustion chamber. To run the compressor, a turbine fan has been placed in the exhaust to covert some of the energy from the high-speed exhaust into mechanical work.

> The Sophia J-850 is a very small jet engine. It weighs 3.08 lb (1.4 kg) and produces 18.7 lb (85 N) of thrust. It is a fully functional turbojet engine designed for instruction and model airplanes.

Conceptually, Figure 4.9 has the components of a turbine engine, although implementation of the components is quite different in practice. Figure 4.10 shows a more realistic drawing of a turbine engine. Basically, it consists of a tube with an inlet (or a *diffuser*), followed by a *compressor* for the air, a *burner* (where the high-pressure air and fuel are mixed and burned), and a *turbine* to power the compressor. At the exhaust end of the turbine there is a *nozzle* to direct the exhaust to give thrust. The three components, compressor, burner, and turbine, are the core of the turbine engine.

Let us go through the different parts of the turbine engine one at a time. The diffuser and nozzle will be discussed a little later because they are not really part of the turbine engine itself but additional parts used to make the turbine engine a jet engine.

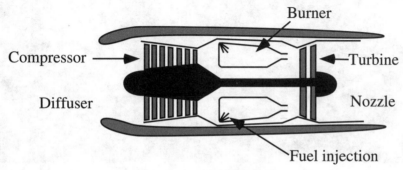

FIGURE 4.10   **More realistic drawing of a turbine engine.**

Compressor

The compressor has two functions in a turbine engine. The first is to act like a one-way valve that prevents the combustion gases from blowing out the front of the engine. The compressor's job is to always push air in one direction—into the burner. The other function implied by its name is to increase the pressure and density of the air and thus the oxygen so that the fuel will burn efficiently. This is particularly important at high altitudes, where there is very little air. The compressor also provides cooling air, as will be discussed later.

Two types of compressors are used in jet engines. Larger engines use *axial-flow compressors*, as shown in Figure 4.11. As the name implies, the air flows down the axis of the compressor, where it is fed directly into the burner. In the figure, only the compressor is shown.

Smaller jet engines are more likely to use *centrifugal compressors*, or *impellers*, as shown in Figure 4.12. Here, the entire engine is shown in cutaway. This type of compressor "pumps" the air to the outer radius of the engine, where it is then redirected into the burners.

Although different in construction, the purpose of both axial-flow and centrifugal compressors is the same—to compress the air. We will now look at the two types in more detail.

FIGURE 4.11   Axial-flow compressor.

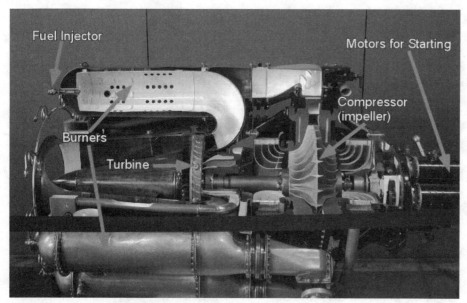

FIGURE 4.12 Centrifugal compressor and engine.

AXIAL-FLOW COMPRESSOR

In an axial-flow compressor, a series of rotating blades pushes the air back and, in so doing, adds energy to that air. Each blade is basically a rotating wing, just like a propeller. There is one fundamental difference between the compressor and the propeller, though. With the compressor, the blades are in a duct, and therefore, the added energy results in an increase in the pressure of the air rather than an increase in the speed of the air. How pressure is produced, rather than speed, is kind of interesting.

The axial-flow compressor is made of rows of blades, consisting of rotating blades followed by stationary blades, as shown in Figure 4.13. A typical row of rotating blades has 30 to 40 blades and is called a *rotor*. Following each rotor is a stationary set of blades called a stator. The rotor's job is to increase the energy of the air and thus its pressure. The *stator* increases the pressure of the air further by slowing it down from the speed at which it left the rotor. The pressure increase across a single stage of rotor-stator combination is fairly low, but multiple stages can produce fairly high compressions with high efficiency. As the pressure increases from stage to stage, the volume of air decreases.

Compressor Blades

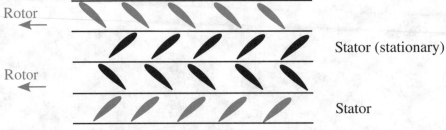

Rotor

Stator (stationary)

Rotor

Stator

FIGURE 4.13 A compressor consists of rotors and stators.

Thus the blades of the rotors and stators become smaller. This can be seen clearly in Figure 4.11.

It is not wise to try to increase the pressure too much across a single stage because this increases the chances of the blades stalling, just like a wing that is trying to produce too much lift. The stall causes the flow to reverse and is referred to as *compressor stall*. Rather than trying to increase the pressure substantially across each stage, multiple stages are used to decrease the pressure gain across each stage. The result is that an entire compressor section of 10 to 12 stages may increase the pressure by a factor of 10 or more.

MULTISTAGE COMPRESSOR

A compressor stage, made up of 10 to 12 rotor-stator stages or a single impeller, can only do so much to compress the air. In principle, for an axial-flow compressor, you could add more rotor-stator stages. However, as the air compresses and slows down, the rotation speed of the shaft becomes too high. The solution to this problem is to add multiple compressors, usually called the *low-pressure* and *high-pressure compressor sections*. Thus most commercial jet engines have multiple concentric shafts. A two-shaft engine is illustrated in Figure 4.14. Figure 4.11 shows a two-stage compressor. The low-pressure compressor has the large blades with seven stages. The high-pressure compressor is on the right and has six stages.

> The engine in Figure 4.12 is a General Electric J-31 *turbojet* engine, which was the first jet engine produced in quantity in the United States. A total of 241 J-31 engines were produced before production ended in 1945.

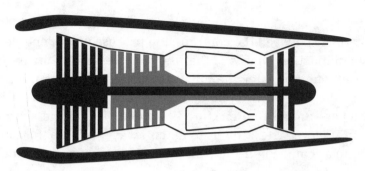

FIGURE 4.14   Two-shaft axial-flow compressor.

Multistage compressors can be built with centrifugal compressors. Some engines have two impellers, whereas others have an axial-flow compressor as the low-pressure compressor, and the impeller used for the high-pressure compressor.

CENTRIFUGAL COMPRESSOR

Centrifugal compressors push the air out radially rather than along the axis of the engine. An *impeller*, shown in Figure 4.12, adds energy by accelerating the air radially. Impellers are popular on smaller engines because a single impeller replaces several rows of blades in an axial-flow compressor, making them much less expensive to build. There is, of course, a drawback to this less-expensive technique. The direction of the air must be turned from heading radially back to flowing along the axis of the engine. This is the source of a significant energy loss. This loss in efficiency is deemed unacceptable in larger jet engines. They also have a greater frontal area than axial-flow compressors, which increases drag.

> In a rush to be the first to fly nonstop from New York to Paris, Charles Lindbergh set a record with a nonstop overnight flight from San Diego to St. Louis.

Centrifugal compressors have remained important since the beginning of jet engines. They produced an engine with a short length, which has made them desirable for small corporate jets and for use in helicopters. They also have proven to be very durable and are less likely to be damaged by foreign objects than axial-flow compressors.

Burner

One difference between the compression process in a jet engine and an internal combustion engine is that the compression of air is continuous in a jet engine. After the air is compressed, fuel is injected and burned in the *burner*, or *combustor*. The burner is merely a kind of firebox where the air-fuel mixture is burned. The burners are clearly visible in Figure 4.12, the top in cutaway and the bottom fully enclosed in a "can." Figure 4.15 shows one of the burners on an engine with an axial-flow compressor.

In most engine designs, the burner is really several burners distributed around the central axis like a series of cans. In Figure 4.15, the air would flow from left to right. After entering the inlet, the air splits, with part going along an outside volume and part going through the combustion chamber. The air flowing down the outside of the combustion chamber is for cooling. The air going to the combustion chamber passes through a plate with holes, which is a *flame stabilizer*. The flame stabilizer enhances the mixing of the fuel with the air and prevents the flame from being extinguished by the rush of air.

For best combustion efficiency, the combustion temperature is kept as high as possible. Current temperatures at the end of the combustion chamber are on the order of 2800°F (1500°C). A typical melting point of

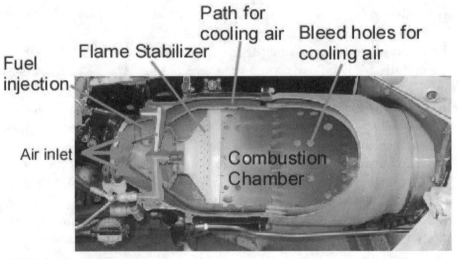

FIGURE 4.15   Axial burner.

steel is 2400°F (1300°C). This temperature is too hot for typical construction materials, so the burner must be cooled. Bleed air is brought in through holes in the wall of the burner to form a thin film covering the inside walls, as seen in the figures. The hot combustion gases do not burn through the combustion chamber walls because of the constantly replenished supply of cool air. Most of the air taken in by a jet engine, as much as 75 percent, is used for cooling, and therefore, typically only about 25 percent of the oxygen is consumed in the burner. This unused oxygen makes afterburners possible, as will be discussed at the end of this chapter.

> A modern turbine blade for a wide-body airliner is roughly 3 in tall and 2 in wide yet produces about 750 hp.

It is in the burner that the energy is given to the air through chemical combustion to produce propulsion. Before the energetic exhaust can be allowed to escape, however, there is some work for it to do. Some of its energy must be extracted to power the compressor. This is done by the turbine following the burner.

Turbine

The turbine looks quite a bit like a single stage of a compressor, only here the first set of blades that follows the burner is fixed and does not rotate. These blades are called the *turbine vanes*. They are followed by a rotating set of *turbine blades* that drive a shaft connected to the compressor. The arrangement is illustrated in Figure 4.16. The purpose of the turbine vanes is to turn the exhaust into the turbine blades. This allows for greater energy transfer to the turbine blades.

A turbine is the reverse of a compressor. The air expands and cools through each turbine stage, and energy is removed from the air. There

Fixed

Rotating ⟶

Turbine vanes

Turbine blades

FIGURE 4.16 Turbine with its vanes and blades.

must be as many turbine sections as there are compressor sections, but not as many stages in each section. Thus a jet engine with two compressors, a low- and high-pressure section, will have two turbines, each powering one of the compressors with a separate shaft. The turbine-shaft-compressor combination is referred to as a *spool*. Most large jet engines are *two-spool engines*, meaning that they have a two-stage compressor driven by a two-stage turbine. This was illustrated in Figure 4.14.

> The radial engines in World War I fighter airplanes actually rotated with the propeller.

Although the exhaust loses some of its energy going through the turbine, and thus becomes cooler, it is still very hot. The first vane-blade stage of the turbine sees temperatures similar to those in the burner, on the order of 2800°F (1500°C). Therefore, this turbine stage requires special cooling, including the film cooling from bleed air as is found in the burner. A turbine blade is shown in Figure 4.17. Notice that it is hollow, which is to allow internal cooling air, and that there are small holes on the surface. These holes allow a cool air pocket to form around the surface of the blade. This pocket is thin but allows the blades to survive the hot temperatures.

The pressure change across the turbine goes from high to low. In the compressor, the pressure goes from low to high. Because of this, unlike the compressor, there is little problem with the turbine blades stalling. Therefore, the pressure change across a turbine stage can be much greater than the pressure change across an axial-flow compressor stage. Thus fewer stages are needed in the turbine than in the com-

FIGURE 4.17 Turbine blade. (Photograph courtesy of NASA.)

pressor. Even with the energy loss, the gas leaving the turbine still has a great deal of energy and can be used for propulsion.

The Turbojet

The simplest form of the jet engine is the turbojet shown in Figure 4.10. Basically, a turbojet is a turbine engine with a diffuser and a nozzle. The diffuser works to "condition" the air before it enters the compressor because the compressor is optimized for a certain air speed. Conditioning by the diffuser helps to bring the air speed at the compressor to its optimal speed regardless of the speed of the airplane. A typical air speed entering a compressor might be half the speed of sound (Mach 0.5). Thus, in a transport that cruises at roughly Mach 0.8, the diffuser slows the air down considerably. When the airplane is standing still at the end of a runway, the diffuser speeds up the air. This might lead you to believe that the diffuser is an active device. Actually, it is the compressor that "demands" how much air must be sucked into the engine. For most jets, the diffuser therefore is passive, making sure that the air is uniform when it hits the compressor at the right speed. At supersonic speeds, it is important for the diffuser to slow the incoming air to subsonic speeds as efficiently as possible.

At the other end of the turbojet is the nozzle, which conditions the exhaust gas as it exits the engine. The ideal situation is to expand the gas back to atmospheric pressure such that it exits at the greatest possible velocity. This gives the greatest thrust. The design of the nozzle depends on the pressures and velocities of the gas after it leaves the turbine.

There are two fundamental problems with the turbojet. First, the turbojet produces thrust with a very high exhaust-gas velocity. We have seen that this produces more wasted power and thus is inefficient. Another problem is that the higher the exhaust-gas velocity, the more noise the engine produces. This noise is unacceptable today, and regulatory noise standards do not permit turbojets to operate at many airports.

Older airplanes, such as the Boeing 707, that used turbojets are now a rare sight at most airports. The Boeing 727 and 737 originally had turbojets, and many of the earlier versions of these airplanes cannot be flown into many airports because of noise restrictions. In the case of the Boeing 737, the airplane has gone through two major redesigns to improve efficiency and noise. Because of its unique requirements, the Concorde used turbojets, making it noisy and a "gas guzzler" by any standard.

Jet Engine Power and Efficiency

The engine power developed by a turbojet, and thus the amount of air drawn into the engine, depends on the amount of fuel injected into the burner and not the speed of the airplane. If more fuel is added, more force is put on the turbine, causing more air to be brought in by the compressor. It is a characteristic of jet engines that engine power and thrust approximately depend only on the throttle setting. Thus an engine can develop full engine power and thrust sitting on the runway or at cruise speed.

Now here comes the interesting part. The power available is the thrust times the speed of the airplane. Therefore, although the engine may be developing full power and thrust as it starts to roll for takeoff, it is producing almost no power for propulsion. This is shown in Figure 4.3. Since the wasted power is the difference between the engine power and the power available, almost all the power is wasted! We have said that the propulsive efficiency is the power available divided by the engine power. This is almost zero at takeoff and increases with speed. One sometimes hears that jet efficiency increases with speed. This is the source of that increase in efficiency. As will be seen in Chapter 6, the fact that a jet's power available increases with speed, whereas that of the piston-driven engine is roughly constant, affects how these planes make climbs and turns.

> The Wright brothers' first engine produced 12 hp and weighed 180 lb.

Turbojets have another advantage over piston-driven airplanes. Because of the design of the diffuser, the amount of air that a turbojet takes in does not depend strongly on altitude. Therefore, turbojets are able to fly high where the parasite power is greatly reduced and still develop full power. This enhances the efficiency of jet airplanes.

The Turbofan

In order to optimize the efficiency of a jet engine, one wants to accelerate a large amount of air at as low a velocity as necessary to produce

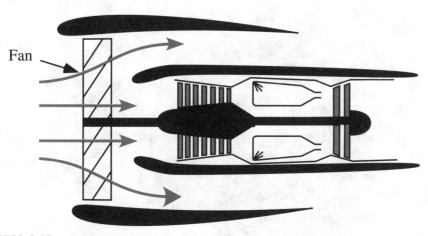

FIGURE 4.18   Turbofan with much of the air bypassing the core.

the needed thrust. The nature of turbojets limits the amount of air that can be processed. The solution has been the introduction of the *turbofan* engine, shown schematically in Figure 4.18. The turbofan engine is designed around a turbine engine, but with much more of the energy produced in the burner being converted into mechanical energy by the turbines. Most of this energy is used to turn a large fan in front of the engine. The fan is very much like a propeller, but with 30 to 40 blades instead of just 2 to 4. The large fan accelerates a large amount of air at a much lower speed than the exhaust of a turbojet producing the same thrust or power. Thus it is much more efficient.

It is important to understand that the air that goes through the fan does not go through the core of the engine but goes around the outside of the core. This can be seen in the drawing of a turbofan in Figure 4.19. Clearly, most of the air that goes through the fan bypasses the core. The ratio of the air that goes around the core to that which goes through the core is called the *bypass ratio*. Typical engines today have bypass ratios of about 8:1, meaning that eight times as much air goes around the core as going through

> Engine power grew rapidly once flight became a reality. By 1910, the French had built an engine that could deliver 177 hp (179.5 kW).

it. Ideally, the exhaust gas and the air from the fan would have the same velocity. In such a situation, with a bypass ratio of 8:1, about 90 percent for the thrust would come from the fan and about 10 percent from the exhaust of the turbojet powering it.

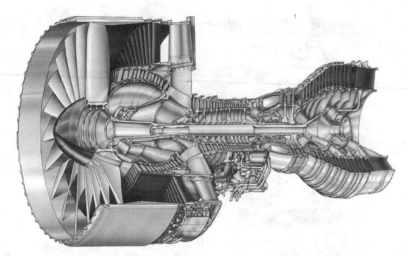

FIGURE 4.19   Turbofan engine. (Courtesy of Pratt and Whitney.)

An additional benefit of the turbofan, and a necessary one, is that the lower exhaust velocity produces less noise. Jet engines today are much quieter than they were 30 years ago. The turbofan also gives engine designers a means for increasing thrust while increasing efficiency. They can increase the mass flow through the engine while decreasing the output velocity. The result of this is that the 115,000-lb-thrust engines on the Boeing 777-300ER have a fan so large that the engine's diameter is within inches of the fuselage diameter of a Boeing 737. One can fit six seats and an aisle in one of these engines, although these would be very uncomfortable seats. Figure 4.20 shows the engine of a Boeing 777 next to a service truck. This gives a feeling for their size.

Unfortunately, there are practical limits to the size of the engines. However, there are always clever engineers out there who find ways to expand these limits. The engines used on the Boeing 777-300ER were unimaginable just three decades ago.

> Modern aircraft piston engines weigh approximately 2 lb/hp.

Turbofans are the engine of choice for commercial transports and business jets. Military jets also use turbofans but must compromise efficiency if the airplane is to be capable of supersonic flight. The huge inlets of the fanjets are a deterrent when it comes to supersonic drag. Military aircraft, which require that the engine have a small frontal area and high thrust, have low bypass ratios, usually significantly less than 2:1.

FIGURE 4.20   Engine of a Boeing 777.

The Turboprop

A turboprop follows the same concept as a fanjet. The excess power not needed to run the core is used to turn a propeller rather than a fan. Figure 4.21 shows a schematic of a turboprop. A propeller cannot be attached directly to this shaft from the turbines because it turns at too high a speed. Therefore, a gearbox or *reduction* gear is used to reduce the rotation speed to the propeller. The shaft powering the gearbox is usually powered by a dedicated turbine not used to drive the compressor.

At lower flight speeds, a propeller is more efficient than a fanjet. Propellers become less efficient at high speeds because of the effects of compressibility, which will be covered in Chapter 5. Special propellers have been designed to operate at these higher speeds, but they have not seen use except on experimental aircraft.

Turboprops are used on smaller commuter aircraft and have found a growing use in the general-aviation and corporate airplane market. An example of a corporate airplane with turboprops is shown in Figure 4.22. The big advantage of turboprops is that they are able to produce greater power than a piston-driven engine of the same weight with much less noise and maintenance. Unfortunately, turboprops are also more expensive than piston engines.

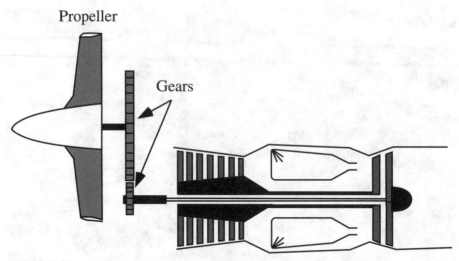

FIGURE 4.21 Turboprop engine.

FIGURE 4.22 Turboprop on an airplane.

Thrust Reversers

A large jet must decelerate when it lands. As with a car, brakes can be used. However, as will be shown in Chapter 6, the energy that the brakes have to absorb is astounding. When a jet lands, it seems as though the pilots have turned the engine around and directed the thrust forward. A jet engine redirects its thrust with a *thrust reverser*. Although it may seem like it, this does not mean that the engine works backwards, that is, blowing gas out the front. What a thrust reverser does is to divert the gas in the jet

> The first jet engines in the German ME-262 had to be overhauled every 10 hours.

exhaust and send it forward. This is illustrated with the *clamshell*-type reverser shown schematically in Figure 4.23. In this figure, the air from the fan is rerouted out through the cowling while the nozzle blocks the core air and forces it forward. The net result is that the engine produces some negative thrust to slow the airplane down. Figure 4.24 shows the thrust reverser on a corporate jet in the stowed and deployed positions.

A thrust reverser can only redirect a portion of the air forward. The net reverse thrust is just a very small component of what it can produce in the forward direction. However, even if the thrust reversers

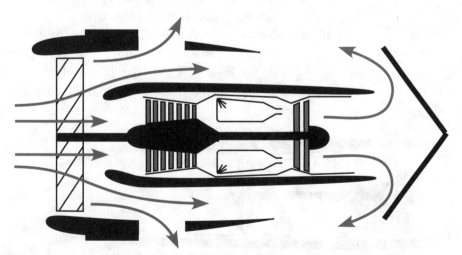

FIGURE 4.23 A thrust reverser partially turns the exhaust forward to produce negative thrust.

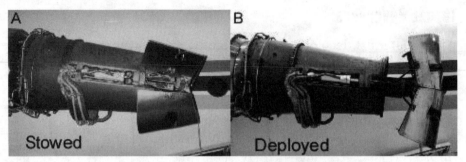

FIGURE 4.24 Clamshell thrust reverser: (a) stowed; (b) deployed.

only canceled the forward thrust, the large engine would produce a great deal of drag. This alone would help to slow down the airplane on landing.

Thrust Vectoring

Sometimes jet nozzles can be pointed in a particular direction other than straight back. This is known as *thrust vectoring*. Figure 4.25 illustrates how a hinged nozzle can redirect the jet exhaust from horizontal to an angle. The idea, similar to that of a thrust reverser, is to redirect the jet in a desired direction. The Harrier is an extreme example of thrust vectoring. The Harrier is able to hover by directing all the thrust down. The exhaust jet is routed to four nozzles (shown in Figure 4.26) that can swivel from horizontal, for forward thrust, to vertical, for hovering. The Harrier

> Today, commercial airplanes have to turn back more often because of inoperable toilets than because of an engine failure.

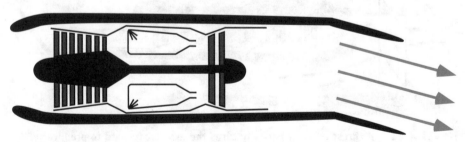

FIGURE 4.25 Thrust vectoring.

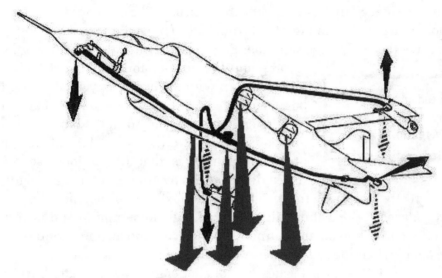

FIGURE 4.26   Thrust vectoring of the Harrier in hover.

also uses the high-pressure gas from the engine for attitude control, as shown. Modern fighter designs, such as the Lockheed-Martin/Boeing F-22 Raptor shown in Figure 2-43, use thrust vectoring to increase maneuverability.

During the Falklands' war between Britain and Argentina, thrust vectoring was found to be extremely effective in combat. Pilots used the thrust vectoring of the Harrier, originally designed for vertical or short takeoff and landing, to improve maneuverability. A pilot in straight-and-level flight could quickly direct some of the thrust down and literally make the airplane hop in the air. This extra maneuverability allowed the Harrier to avoid air-to-air and surface-to-air missiles at the last minute by hopping out of the way.

Afterburners

Military fighters and interceptors sometimes need extra power, called *military power*. One solution is to install a larger engine. However, a larger engine weighs more and thus is not a practical choice. Instead, some military engines *augment* their thrust. The principle is to add more power to the air after the turbine has removed much of the power to drive the compressor and fan. Fuel injectors are added between the turbine and the nozzle, and they inject fuel to mix and burn with the excess oxygen. It should be remembered that little of the oxygen taken in by a jet engine is consumed in the burner. This device is called an *afterburner*. The advantage of an afterburner is that it can increase the thrust of the engine by 50 percent or more without greatly increasing the weight or complexity of the engine. The disadvantages are that it entails very high fuel consumption and is very inefficient. When an afterburner is being used, the engine is said to be *wet*, and when the afterburner is off, the engine is *dry*.

> Engine noise from today's 747-400 is half of what it was on the original 747s delivered in 1970.

Figure 4.27 shows a schematic of a jet engine with an afterburner. After the turbine, a ring of fuel injectors has been added. A tube and flame holder and a variable nozzle also have been added. The variable nozzle is an important component of an afterburner. The addition of a large amount of burning fuel increases both the mass flow and velocity of the exhaust. The nozzle must increase the area of the outlet, or the pressure would increase at the downstream side of the turbine. This could cause a compressor stall or a fan surge on a turbofan. Figure 4.28 shows the nozzles of an FA-18 Hornet. The left nozzle is set for dry operation, whereas the right is adjusted for wet operation. The difference in nozzle throat area is significant. Figure 4.29 shows the nozzle of an SR-71 Blackbird. The four white rings inside are the fuel injectors for the afterburner.

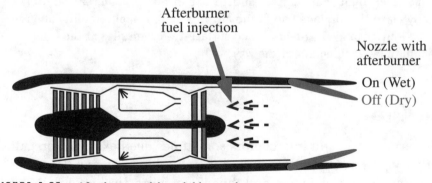

FIGURE 4.27   Afterburner with variable nozzle.

FIGURE 4.28 Nozzles on an FA-18 Hornet, left set for cruise and right for after-burner on.

Visually, an afterburner is impressive. Because the combustion is taking place just before the nozzle, the flame actually extends through the nozzle and out the rear of the engine. Figure 4.30 shows an FA-18 Hornet taking off from an aircraft carrier at night on afterburners.

Because of the high fuel consumption, afterburners are used only for short periods. The one outstanding exception is the SR-71 Black-

FIGURE 4.29 Afterburner of an SR-71. White rings are the fuel injectors.

FIGURE 4.30 FA-18 Hornet taking off at night on afterburners.

bird. Because of its high speeds and altitudes of operation, it can operate with reasonable efficiency on afterburners. Figure 4.31 shows an SR-71 Blackbird on afterburners. The flame is easily visible. The bright spots are called *shock diamonds*. These are caused by supersonic exhaust, which produces shock waves that will be discussed in Chapter 5. The apparent oscillations are due to shock interactions caused by differences between the exhaust pressure and the ambient pressure.

Wrapping It Up

Aircraft propulsion systems involve the same physics as wings. In the case of the propeller, it is nothing more than a rotating wing. The propeller accelerates air back, which pushes the airplane forward. The same is true of a jet engine. The jet accelerates air back to produce thrust. The efficiency of the engine is determined by a compromise between the amount of air flowing through the engine per second and the jet velocity. For subsonic speeds, the fanjet maximizes the airflow through the engine and is the most efficient for transports traveling just below the speed of sound. Fanjets are fuel efficient and relatively quiet. At supersonic speeds, the turbojet is more efficient.

> The SR-71 was designated the RS-71 until President Johnson accidentally reversed the letters in a public announcement. Rather than embarrass the president, the designation was changed.

FIGURE 4.31   The SR-71 on afterburners. (Photograph courtesy of NASA.)

Next, we will look at high-speed flight. There are profound differences in flight at or above the speed of sound, where information cannot be communicated forward and the compressibility of air becomes very significant.

High-Speed Flight

Supersonic fighters, commercial transports, and bombers fly at speeds greater than most small airplanes. The speeds at which they fly introduce some additional physics that are not present in low-speed flight. In particular, the compressibility of the air becomes significant. We have said in examining lift that air is to be considered an incompressible fluid because of the low pressures involved with low-speed flight. However, air *is* compressible. All fluids are compressible on some level, even water. When air is compressed, its density changes.

> Hypersonic vehicles do not have sharp noses and wings because there must be enough material to absorb the heat. Sharp objects will burn off at high speeds.

Figure 5.1 shows a fighter flying at just below the speed of sound. The air flowing over and under the wings has expanded, lowering its density and temperature, which causes water vapor to condense. In high-speed flight, one must understand where, when, and how the density of the air changes.

Mach Number

Low-speed flight is *subsonic* flight. High-speed flight can be broken into three basic categories: *transonic*, *supersonic*, and *hypersonic*. As their names imply, the categories are related to the speed of sound

FIGURE 5.1   Transonic F-18 with shock wave over wing. (Photograph courtesy of the U.S. Air Force.)

(sonic). As stated earlier, the Mach number is the airplane's speed in units of the speed of sound. At Mach 1, the plane is going at exactly the speed of sound. *Subsonic* refers to speeds below Mach 1. *Transonic* refers to speeds near Mach 1. Commercial transports, most military transports, bombers, and business jets fly at transonic speeds. *Supersonic* refers to speeds above Mach 1 and is usually left to fighters and interceptors for short bursts. *Hypersonic* refers to speeds of high Mach numbers. At present, the only hypersonic vehicle is the space shuttle during reentry.

Mach number also relates the apparent speed of the air as seen by the airplane. The airplane travels at a single Mach number with respect to the air at a distance. The airflow around the airplane is traveling at different speeds and thus has a different local Mach number. In transonic flight, significant parts of the airflow, relative to the aircraft, are both subsonic and supersonic. In the case of supersonic flight, virtually all the air relative to the airplane is supersonic. Finally, in hypersonic flight, the vehicle is going so fast that certain additional physical phenomena creep in. This will be discussed in a later section.

Why is the compressibility of the air important? The consequences of compressibility may confuse our intuition about the behavior of air.

Remember: Lift Is a Reaction Force

High-speed flight, like low-speed flight, requires that air be diverted to create lift. This basic principle does not change with the introduction of compressibility. Ultimately, as an airplane flies overhead, it diverts air down as it passes, regardless of its speed. However, there are some changes as to how this occurs at high speeds. Recall that in low-speed flight there is upwash in front of the wing owing to circulation. As speed increases, upwash begins to disappear. Upwash was possible at low speeds because air transfers information at the speed of sound around the wing. Thus the air is able to transfer information in front of the wing, and thus the air is able to adjust for the oncoming wing. There is upwash, and the airflow separates before the arrival of the wing. As the wing moves faster, there is less time for the air ahead to move out of the way of the wing. Once the airplane becomes supersonic, upwash ceases. Also, an almost instantaneous compression wave known as a *shock wave* forms in front of the wing. Understanding shock waves and their drag and power consequences is somewhat complicated. It helps to first understand some of the fundamental properties of supersonic air. The next few sections will present a basic primer on supersonic airflow.

> The American X-15 reached a speed of 4534 mi/h (7295 km/h), or Mach 6.72.

Compressible Air

While the forces involved in flight at low speeds can be quite large, the pressure changes are relatively low, so the air is often considered an incompressible fluid. If we were to flow such a fluid through a pipe with a constriction in it, what would we see? The answer can be found in the discussion of the venturi in Appendix A. As the fluid comes to the constriction, the velocity increases and static pressure (measured perpendicular to the flow) decreases. Thus, as velocity goes up, static pressure goes down, and because the fluid is incompressible, the density and temperature remain the same. In general, for an incompressible fluid, the velocity and static pressure change in opposite directions, and the density and temperature remain constant.

As the speed of the air increases, the pressures become significant, and the compressibility of the air must be considered. Now, the density of the air can change, and thus its volume no longer remains constant. At this point, for example, in a constriction, the velocity decreases and pressure increases, whereas density and temperature also increase.

As we will discuss in greater detail later in this chapter when we examine supersonic wind tunnels, there is another big difference between the behavior of supersonic airflow and subsonic air. The difference is that supersonic air cannot communicate upstream (forward in the rest frame of the wing) and can only communicate a very short distance directly above a point on a wing. This will have a profound effect on the behavior of air flowing over a wing at transonic or supersonic speeds.

There is an additional element in supersonic aerodynamics that does not exist in subsonic aerodynamics. This is the formation of shock waves.

> Traffic patterns behave like supersonic airflow. In a constriction, when a highway lane is closed, for example, speed goes down, and the density of the cars increases. Looking at the faces of the unfortunate drivers stuck in the traffic jam, one might infer that pressure and temperature also have increased!

Shock Waves

Compression in air can happen over such a small distance that it forms a shock front or *shock wave*. In supersonic flight, a shock wave occurs when air must change speed and/or direction suddenly. Figure 5.2 shows a shock wave on a Space Shuttle model in a supersonic wind tunnel.

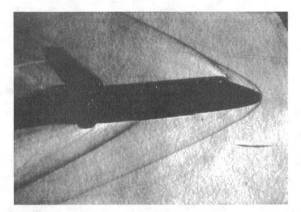

FIGURE 5.2   Shock wave on a model of the Space Shuttle.

There are two types of shock waves of interest involved with flight: *normal* (meaning perpendicular) shock waves and *oblique* (at an angle) shock waves. Normal shock waves are perpendicular to the direction of flight and are seen primarily on the surfaces of transonic wings. They are caused by an abrupt change in air density and pressure. Figure 5.3 shows what happens across a normal shock wave. Before the shock wave, the air is traveling at greater than Mach 1. Behind a normal shock wave, the air is subsonic, and its density has increased.

> Although the first supersonic flight did not occur until 1947, Ernst Mach photographed the shock waves on a supersonic bullet in 1887.

Oblique shock waves are formed at an angle with respect to the oncoming air and occur when supersonic air must be turned. Because a supersonic airplane is traveling so fast, the air has no chance to move out of the way as it does at subsonic speeds. Therefore, the moment the air hits the leading edge of the wing, it must turn. The air turns almost instantaneously and forms the oblique shock wave. However, the shock wave forms at a given angle depending on the angle it must be turned and the speed at which the object is moving. Figure 5.4 illustrates an oblique shock wave. As with a normal shock wave, the air density increases and its velocity decreases across an oblique shock wave. The changes are not enough, however, for the air to become subsonic, as in a normal shock wave. Therefore, the air behind an oblique shock wave remains supersonic relative to the aircraft, with some rare exceptions to this rule.

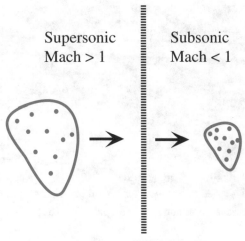

Supersonic
Mach > 1

Subsonic
Mach < 1

Normal Shock Wave

FIGURE 5.3 Density and Mach number change across a normal shock wave.

All supersonic objects create shock waves. Normal shock waves cause a higher change in air density than oblique shock waves. The greater the change in density across a shock wave, the greater is the energy loss to compression, and the greater is the drag. Therefore, supersonic aircraft are designed to avoid producing normal shock waves. This is accomplished by making the nose and wing leading edge sharp. Blunt noses lead to energy-consuming *bow shocks*, which

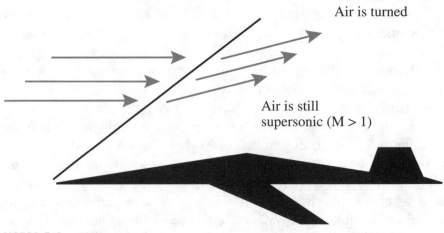

Air is turned

Air is still
supersonic (M > 1)

FIGURE 5.4 Oblique shock wave.

are a combination of a normal shock wave on the very nose joined to an oblique shock wave a little ways back. Bow shocks are avoided by putting sharp noses on supersonic airplanes.

Shock waves in air have known density, pressure, and velocity jumps. For a given shock angle, all these properties can be found in tables. Thus, for a given airplane geometry, the shock angles and pressures are easily determined. Supersonic flight is actually easier to analyze than subsonic flight. In low-speed aerodynamics, engineers must rely on complicated equations to solve for the pressures over a vehicle. In supersonic aerodynamics, an engineer can use published tables. However, being able to compute the pressures on the vehicle more easily does not translate into making the problem of supersonic flight easier. The penalty of supersonic flight is wave drag.

> A World War II DC-3 lost a wing from a bomb while on the ground. The only available replacement was a DC-2 wing, which was 5 ft shorter and designed for a much smaller load. The wing was attached, and the airplane, dubbed a DC-2½, flew to safety.

Wave Drag and Power

As discussed in Chapter 1, the faster an airplane goes, the greater is the amount of air that is diverted. Thus a smaller vertical velocity of the downwash is needed, and the induced power loss is smaller for the same load. Therefore, the induced power is small for very fast airplanes. The downside is that the parasite power goes as the speed cubed. Now, there is an additional demand for power at high speeds. The demand is to overcome *wave drag*.

Shock waves travel with the aircraft. Before supersonic flight over land was banned, *sonic booms* (a sound like an explosion heard by those on the ground) were a frequent occurrence around military bases. The shock wave is a persistent phenomenon that travels along with the aircraft and extends for miles. This means that the airplane is not only doing work locally to change the airflow but is also affecting the air miles away! This results in an increase in wave drag and an increase in the required power. The extra power needed to overcome wave drag is one factor that makes supersonic flight so difficult.

Wave drag is more complicated than induced drag and parasite drag. Unlike the latter two contributions to drag, wave drag is not a simple function of speed but is rather a complicated function of Mach number. For example, if you double the Mach number, and thus the speed, the power needed to overcome wave drag may increase by less than a factor of 3, whereas the parasite power would increase by a factor of 8. Since drag is power divided by speed, the wave drag has increased by less than 50 percent in this example, whereas the parasite drag has increased by a factor of 4.

> A supersonic airplane flying at 60,000 ft can produce a sonic boom that reaches about 30 miles to either side of the flight path.

So why does the wave drag decrease so slowly? The drag caused by oblique shock waves depends on the angle the shock wave makes with the direction of the airplane's travel. The more perpendicular the shock wave, the greater is the drag. As the Mach number increases, the angle of the shock wave decreases. This is illustrated in Figure 5.5. Thus there is a counteracting effect of increasing Mach number. The shock strength increases with Mach number, but because the shock angle is smaller, the wave drag does not increase very fast. The angle of the shock wave is a function of the body angle. Therefore, supersonic aircraft have very sharp noses and leading edges so that an oblique shock wave is produced at as small an angle as possible.

An additional consideration in terms of wave drag is that it is a function of air density. The greater the density of the air, the higher is the wave drag, and thus the greater is the power required to overcome it. Thus aircraft do not fly supersonically at low altitudes.

Transonic Flight

Commercial transports fly in the range of Mach 0.8 to 0.86, just below the speed of sound. This speed is not chosen arbitrarily. It is based on

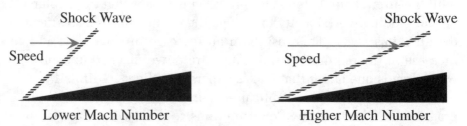

FIGURE 5.5 As the Mach number increases, the angle of the shock wave decreases.

the presence of wave drag. However, if the airplane is flying at a speed less than the speed of sound, how can there be wave drag?

A wing diverts air down. In bending the air down, it creates lower pressure on the upper surface of the wing, which causes the air to accelerate. At speeds approaching the speed of sound, the air that is accelerated over the top of the wing becomes locally supersonic.

When air flows over the top of a subsonic wing, it accelerates to the point of greatest curvature of the air. At this point, the pressure is the lowest, and the speed of the air is greatest. From this point to the trailing edge of the wing, the air speed decreases and the pressure increases in order to match the pressure of the air at the trailing edge. This is the *trailing-edge condition*.

The picture is quite different for the air flowing over the top of a transonic wing. The air accelerates as before, but by the time it reaches the point of maximum curvature, it is traveling at greater than Mach 1. As the air continues to bend, because it is traveling faster than the communication speed of air, it is not able to effectively pull air down from above. Thus the density is substantially reduced, causing the pressure to continue to go down and the velocity to increase. This situation leaves the problem of how to meet the trailing-edge condition. The result is the formation of a normal shock wave, as shown in Figure 5.6. As this shock wave propagates, the pressure and density of the air increase abruptly, and the velocity of the air goes below Mach 1. After the shock wave, the air can slow down further, and the pressure continues to increase to meet the trailing-edge condition. After the shock wave, the flow separates from the wing, causing an increase in form drag.

> **Pilots of the X-15 flew so high that they were given astronaut wings.**

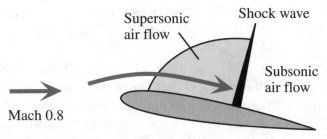

Supersonic air flow

Shock wave

Subsonic air flow

Mach 0.8

FIGURE 5.6 A transonic airfoil with a shock wave.

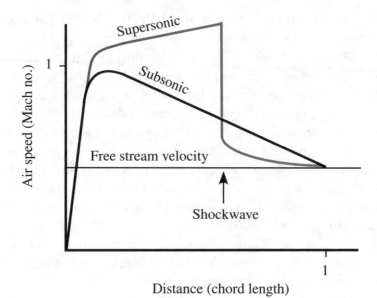

FIGURE 5.7 Airspeed over a transonic and a subsonic wing.

Figure 5.7 shows the air speed, in units of Mach number, for a subsonic and a transonic wing. The subsonic airfoil is traveling at a speed just below the *critical Mach number* such that the air never reaches supersonic speed. The transonic airfoil is just above the critical Mach number, so the air becomes supersonic. The Mach number of the subsonic airfoil decreases after the peak, whereas the Mach number of the transonic airfoil increases until the shock wave. Figure 2.19 shows a winglet on the end of a transonic wing. The winglet only covers the end of the wing after the shock wave.

So how does the wing know where to put the normal shock wave? Let us first assume that the normal shock wave forms at the trailing edge in order to meet the trailing-edge condition. What we would find is that the pressure difference across the shock wave results in a pressure that is higher than the trailing-edge condition. The higher pressure behind the shock wave has a greater forward force than the drag caused by the shock wave. Thus the shock wave will move forward on the wing. As it does, the pressure difference decreases until the wave moves to a place on the wing where the force from the pressure difference just equals the force owing to the drag of the shock wave. If the airplane were now to increase its speed, the drag of the shock wave would increase, and the shock wave would move back toward the trailing edge. At some speed, the normal shock wave will in fact reach the trailing edge.

Look again at the fighter flying at transonic speed in Figure 5.1. In the region where the air is supersonic, density and temperature are decreasing. At a point before the normal shock wave, the air has cooled enough to cause condensation, producing the cone of fog above and below the wing. The backside of the cone is a flat surface. This is the location of the normal shock wave, where the pressure and temperature increase and the condensation disappears. One may ask why there is a normal shock wave on the bottom of the wing. The fighter has almost symmetric wings, and since the angle of attack is so small at transonic speeds, there is a reduction in pressure and acceleration of the air on both the top and the bottom of the wing. It is just that the acceleration and reduction in pressure are not as great on the bottom. Commercial jets that fly transonic speeds are designed so that the normal shock wave forms only on the top of the wing.

> A Concorde "lapped" a Boeing 747 on June 17, 1974. The Concorde left Boston and flew to Paris and back in less time than it took the Boeing 747 to fly one trip from Paris to Boston.

> In the early attempts at breaking the sound barrier, the jump in the center of lift caused the wing to pitch down. This reduced the acceleration of the air over the wing, causing it to become subsonic again. The center of lift then would jump forward on the wing, and the wing would pitch up, becoming transonic. This process would repeat itself rapidly until the wings broke off. The X-1 made it through this transition because of a better wing design that made it strong enough to enter supersonic flight.

In transonic and supersonic flight, the velocity of the air over the wing continues to increase until the normal shock wave is reached. Because of this, the center of lift is farther back on the wing than in subsonic flight. For a typical wing in subsonic flight, the center of lift is about ¼ chord length back from the leading edge of the wing. This means that the wing produces 50 percent of the lift by that point. The moment the wing becomes transonic, the center of lift moves farther back. As the speed increases further, the center of lift continues to move back. At very high-speed flight, the center of lift can move as far as ½ chord length back from the leading edge.

As the airplane flies faster and starts to go transonic, the shock wave begins to form, with the airflow separating from the wing behind the shock wave owing to the pressure increase. The result is a rapid increase in wave and form drag, and thus more power is required. The rapid rise in drag and in required power as one approaches Mach 1 is called the *Mach 1 drag rise* and is shown in Figure 5.8. The design of the wing and body of the airplane is optimized to operate at the "elbow" (or sharp increase) in the figure. Note that the power has a local peak at Mach 1 and then drops off before continuing to rise again. The dropoff is due to the shock wave moving back on the wing with increased speed, causing less of the air to separate from the wing. This reduces the form drag. The smooth continuum that the peak sits on is the parasite power of the aircraft.

The movement of the shock wave and of the air separation is illustrated in Figure 5.9. The three airfoils represent three different flight Mach numbers. In airfoil A, the flight Mach number is only enough to accelerate the air to supersonic speed on the top of the airfoil. In airfoil B, the flight Mach number is such that the airflow becomes supersonic on both the top and bottom surfaces, although to a lesser speed on the

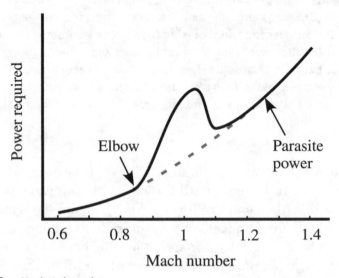

FIGURE 5.8   Mach 1 drag rise.

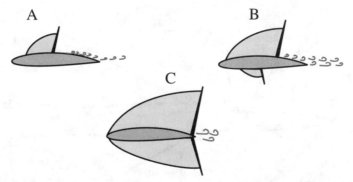

FIGURE 5.9   Shock waves on a transonic airfoil at three different Mach numbers.

bottom surface. Airfoil C shows a Mach number near Mach 1, where the shock waves on both the top and bottom surfaces have become strong enough to move to the trailing edge and produce a substantial drag-inducing wake. This is the situation of the jet fighter in Figure 5.1.

Some early supersonic airplanes had to enter a dive to transition from subsonic to supersonic flight. Only after reaching supersonic flight could the airplane level off and maintain supersonic flight with the available power.

Wing Sweep

Chapter 2 discussed wing sweep. In transonic and supersonic flight, a swept wing is necessary to reduce wave drag. In 1935, a group of top aeronautical scientists from around the world gathered in Rome and showed the results of high-speed analyses and wind tunnel experiments. One German result went largely unnoticed. This result was that a swept wing reduced drag at high speeds.

The reduced drag of a swept wing results from reducing the effective Mach number of the wing. The *effective Mach number* is the Mach number the wing sees perpendicular to the leading edge, as illustrated in Figure 5.10. A

> Supersonic flight is actually easier to analyze than subsonic flight.

nonswept wing will experience the sharp rise in the power requirement (see Figure 5.8) at a lower Mach number than a swept wing. At supersonic speeds, wing sweep also helps to reduce the strength of the oblique shock wave from the leading edge of the wing.

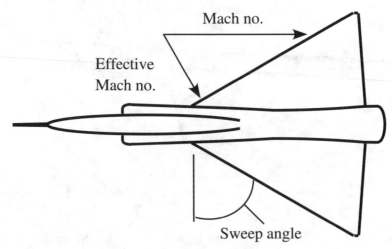

Effective
Mach no.

Mach no.

Sweep angle

FIGURE 5.10 Effective Mach number on a swept wing.

The increase in drag at Mach 1 was not understood at the time of the Rome conference. The discovery of the airflow separation led to better airfoil designs for transonic and supersonic flight. The wake owing to separation of the air is indicated on all three of the wings in Figure 5.9.

For supersonic aircraft, there is a simple relationship between the angle of sweep and the design supersonic speed of the airplane. This is illustrated in Figure 5.11, which shows the sweep angle as a function of Mach number. The purpose of sweep for supersonic aircraft is to keep the effective Mach number at the wing leading edge at or below Mach 1. It does not take sophisticated military intelligence to determine the supersonic operating conditions of an adversary's airplane. All one has to do is look at the sweep angle. For example, a Mach 2 airplane will have a sweep angle of 60 degrees.

> Eddie Rickenbacker, the U.S. "ace of aces" in World War I, was a famous auto racer before the war and owned the Indianapolis Speedway after the war. He also owned and built up Eastern Airlines.

Area Rule

The amount of wave drag on supersonic aircraft is a function of the size of the aircraft. To illustrate this point, imagine throwing a small pebble into a still pond. Small waves propagate away from the entry

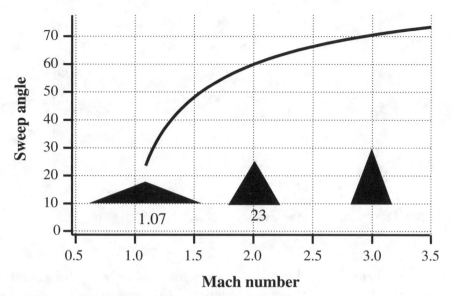

FIGURE 5.11 Sweep angle as a function of cruise Mach number.

point of the pebble. Now repeat the experiment with a big rock. The waves naturally are much larger. The larger waves have more energy than the smaller waves from the pebble. In supersonic flight, the larger the disturbance, the more energy goes into the waves. Thus supersonic aircraft should be thin and sleek.

Some sophisticated analysis performed in the 1940s and 1950s showed that wave drag is proportional to the cross-sectional area (area seen looking at the airplane from the front) of the airplane. At the nose of the airplane, the effect of wave drag grows as the fuselage cross section increases to include the canopy, etc. However, when the wing is reached, the cross-sectional area grows dramatically, which causes a large increase in drag and power required. The solution is to put a "waist" into the fuselage to maintain a constant cross-sectional area. That is, a slice through the wing and fuselage will have the same area as a slice through just the fuselage either before or after the wing. Maintaining of a constant cross section to reduce wave drag has become known as the *area rule*. Figure 5.12 illustrates the area rule. In the figure, fuselage A has less wave drag than fuselage B because it has a waist to compensate for the wings. Fuselage C has the same wave drag as fuselage A but at the cost of a small fuselage everywhere.

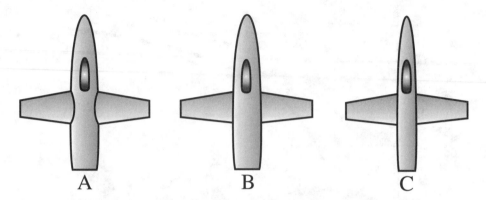

FIGURE 5.12 Illustration of area rule.

The area rule is used in the design of most modern fighter aircraft. A notable airplane with an obvious use of the area rule is the T-38 Talon, shown in Figure 5.13.

In the preceding discussion, a simplification was made. To illustrate the area rule, we used the cross-sectional area as seen from the front, that is, the area of a slice perpendicular to the axis of the airplane. In reality, the cross-sectional area must be held constant along a slice at an angle that is a function of the airplane's design Mach number. Illustrating this complicated result is beyond the scope of this book.

FIGURE 5.13 T-38 Talons with fuselage waist. (Photograph courtesy of the U.S. Air Force.)

Hypersonic Flight

When the Mach number gets high (above about Mach 5), several things happen. First, the aerodynamics become independent of Mach number. This means that for analysis purposes, simple assumptions can be made, and in fact, the analysis of idealized hypersonic flight is the easiest of all aerodynamic analyses. Figure 5.14 shows an artist's conception of a hypersonic airplane. The only known existing hypersonic aircraft is the Space Shuttle during reentry. The X-15, shown in Figure 5.15, explored hypersonic flight in the 1960s, reaching an unofficial speed of Mach 6.7. After the record-making flight, the airplane was retired owing to heat damage from the flight. Recently, the X-43A achieved Mach 10 flight with an unmanned vehicle using an air-breathing engine. The aircraft was not designed to be recovered after its flight.

> Seismic recorders in California record earth movement resulting from the shock waves of the Space Shuttle as it approaches NASA Dryden (Edwards Air Force Base) for landing.

The second change that occurs in hypersonic flight is that the energy transfer of the fast vehicle to the surrounding air becomes so great that the air chemistry begins to change. Oxygen and nitrogen molecules begin absorbing energy and break up, or dissociate. This has many implications for the design of hypersonic aircraft. First, the changing air composition can affect the aerodynamics. This was observed in the Space Shuttle, where the directional and attitude stability was predicted to be higher than it actually was. Fortunately,

FIGURE 5.14 Depiction of Hyper-X hypersonic vehicle. (Illustration courtesy of NASA.)

FIGURE 5.15 Rocket-propelled X-15. (Photograph courtesy of NASA and the U.S. Air Force.)

enough of a safety margin was designed into the spacecraft that this shortcoming was not catastrophic. The second major implication is on skin heating, which will be discussed in the next section.

Skin Heating

Thermal protection requirements of hypersonic aircraft are also affected by the dissociation of the air. Vehicles traveling at high Mach numbers will experience extremely hot gases. Some of this is due to temperature increases across shock waves, and some is from skin friction. The high-temperature air will burn right through any normal material. The Space Shuttle uses ceramic tiles for thermal protection. The dissociation of air molecules actually helps to keep the vehicle cooler. It takes energy from heat to break the chemical bonds of the molecules. Thus heat energy is converted to chemical energy, and the surface temperatures do not get as high as otherwise would be predicted. However, they still get very hot, so the surface must be protected.

> The world altitude record for an airplane was set on August 22, 1963, at 354,200 ft (107,960 m) by an X-15.

Extremely high-speed flight is experienced during reentry to the atmosphere. The Space Shuttle and the Apollo and Soyuz capsules all had to endure very high heat on reentry. When the Space Shuttle first hits the atmosphere, it is traveling at approximately 14,000 mi/h

Sir Isaac Newton invented hypersonic analysis using trigonometry. In 1687, Newton was asked to produce better aerodynamic shapes for cannon artillery projectiles. With no previous work in the field, Newton had to invent his own theories. He reasoned that particles of air collide with the surface and then follow the surface after the collision. The results of his analysis are now known as *Newton's sine-squared law*, which has proven to be a very good predictive rule for hypersonic aerodynamics. However, the first hypersonic flight took almost another three centuries to realize. Of course, Newton did not understand the implications to hypersonic flight. He was trying to solve a much lower-speed problem, for which his theory was flawed.

(23,000 km/h). The thin air that slams into the nose of the Space Shuttle converts kinetic energy to heat. In theory, the air that impacts the nose of the Space Shuttle will reach over 36,000°F (20,000°C), which is about four times the temperature of the sun! Can this really happen?

When the air reaches high temperatures, it goes through complex changes. As mentioned, some of the energy of impact goes into breaking chemical bonds rather than creating heat. Oxygen dissociates and ionizes. The impact is so great, in fact, that the ionized gas that develops around a vehicle reentering the atmosphere prevents radio communication with the outside. This is what is known as the *reentry blackout* experienced by all spacecraft since the first successful atmospheric reentry (the Russian *Sputnik 5*). Rather than having skin temperatures reaching 36,000°F (20,000°C), the temperature is closer to one-fourth that value, but still sunlike temperatures.

> Test pilots took a Boeing 747 to Mach 0.99 during tests.

With these high temperatures, most things that reenter the atmosphere burn up. "Shooting stars," or meteors, are nothing more than small meteorites that burn up as they skip across the earth's atmosphere. Old satellites usually never get as far as the ground before completely burning. So how do the Space Shuttle and the command capsules reenter at equivalent speeds?

The Apollo command modules had a special carbon-based surface on their base. This surface burned off slowly as the craft reentered the atmosphere. The burning heat shield resulted in two effects. The first was that the burning consumes energy, and thus heat, from the air. The second is that the by-products were swept away, taking heat along with them. The astronauts thus were kept cool behind this heat shield. This form of skin cooling is called *ablation.* The problem with ablation is that it is not reusable.

Special tiles were designed for the Space Shuttle that are extremely poor heat conductors. The tiles conduct heat very slowly. When the surfaces of the tiles reach the high reentry temperature, the tiles radiate heat out to maintain a constant surface temperature. However, the longer the Space Shuttle experiences the heat, the deeper the heat will penetrate into the tile. Thus the tiles must be thick enough to prevent the heat from reaching the aluminum skin before the heat load is removed on landing.

Note that airplanes that fly at slower supersonic speeds experience substantial skin heating. Figure 5.16 shows the air temperature as a function of airplane speed from the kinetic energy of the air and not including skin friction. Now if we add friction heating, at about Mach 4, objects glow red. The Concorde fuselage extended as much as 10 in

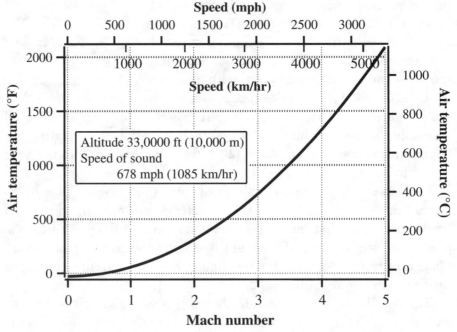

FIGURE 5.16   Air temperature from kinetic energy as a function of airplane speed.

(25 cm) owing to thermal expansion in cruise. The SR-71 experiences such heat that the top of the wing is corrugated on the ground. The thermal expansion of this surface in flight will remove the corrugations to produce a smooth surface. The SR-71 fuel tanks also must accommodate thermal expansion. On the ground, when the SR-71 is cold, the fuel tanks have gaps that leak fuel. By the time the SR-71 has reached cruise, the fuel tanks have expanded to close all the gaps.

Wrapping It Up

Newton's laws are still applicable in describing lift on high-speed airplanes. However, high-speed flight involves some additional physics such as shock waves, wave drag, and high temperatures. The shock waves arise because the vehicle travels faster than the air can pass information—the speed of sound. Next, we will look at the overall performance of airplanes and consider tradeoffs that occur when designing for specific missions.

Airplane Performance

Airplane performance has increased spectacularly in the first 100 years of flight. Improvements in wing design, engine performance, and structural design have led to increased range, speed, endurance, and number of passengers carried. Take, for example, regular air travel from New York to Los Angeles since 1930. Figure 6.1 shows how transcontinental air transportation has changed. The trip has gone from over 2 days to make the trip to about 6 hours. The major reductions in time to cross the country result from improvements in airplane performance. For example, the step at 1930, which represents a 20 percent reduction in travel time, occurred when airplanes could fly at night. Before that time, passengers would transfer to trains when it got dark and resume air travel further along the line in the morning. Although not shown in the figure, there was a slight increase in travel time in 1973 owing to

> In 1911, "Cal" Rogers made the first transcontinental flight of the United States, taking 49 days to fly from New York to Los Angeles.

the oil embargo and a shift to more efficient cruise speeds. The oil embargo caused wings to be redesigned for slightly slower flight in order to conserve fuel.

Figure 6.2 shows the increase in the number of passengers that could be carried by a passenger airplane through the years. In 1927, the Ford Tri-Motor could carry only 11 passengers. The Airbus A380

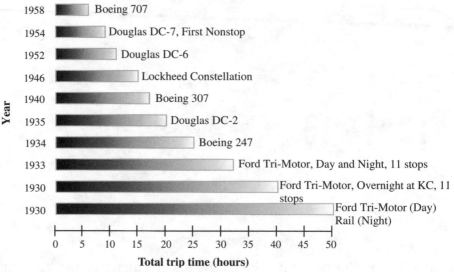

FIGURE 6.1 Transcontinental travel time, 1930–1958.

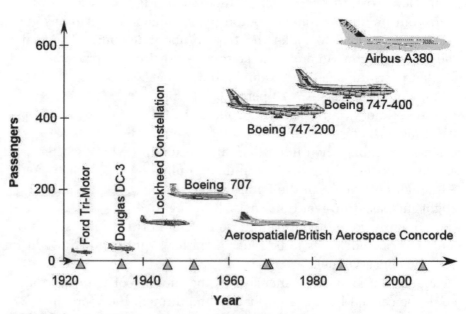

FIGURE 6.2 Increase in number of passengers carried.

eventually will carry 800 passengers, 72 times the Ford Tri-Motor's capacity, although the current version is designed for only 660. Ironically, at the time of this writing, the carriers are using the "luxury" carrot for first class. They are doing this by reducing seating capacity, so the Airbus A380s currently flying seat only about 500 passengers.

In the sections that follow we will discuss the performance of an airplane in powered flight from takeoff to landing. Before that, though, we will prepare you by introducing the lift-to-drag ratio, the glide, and indicated airspeed.

Lift-to-Drag Ratio

There are several parameters that are fundamental to understanding performance. These parameters do not necessarily improve our understanding of how or why airplanes fly, but they are useful aids to understanding airplane performance. An important aerodynamic parameter is the *lift-to-drag ratio*, often referred to as "*L* over *D*" and written *L/D*. Anyone interested in airplanes has likely heard these words at one time or another. *L/D* combines lift and drag into a single number that can be thought of as the airplane's efficiency for flight. Since lift and drag are both forces, *L/D* has no dimensions, which means that it is just a number with no units. A higher value of *L/D* means that the airplane is producing lift more efficiently.

In still air, the *L/D* is the glide ratio, which will be discussed in more detail below. You can determine the *L/D* of a toy balsa-wood glider by measuring its glide ratio, which is the ratio of the launch height to the distance flown (Figure 6.3). This ratio will be the *L/D* of the glider. It is unlikely that this value of *L/D* will be the maximum value, but rather one reflecting how the trim is set for the glider.

> In 1921, Bessie Coleman was the first African-American woman to receive a pilot certificate. She had to go to France to obtain it because she was not allowed to do so in the United States.

There are two ways to look at *L/D*. If you were an engineer designing an airplane, you would have influence over both lift and drag. Once the maximum *L/D* is determined, as well as angle of attack at which it occurs, other performance parameters of the airplane begin to fall out. We shall see this in the discussion of ceiling, range, endurance, climb, and turns. For a pilot, the lift in straight-and-level

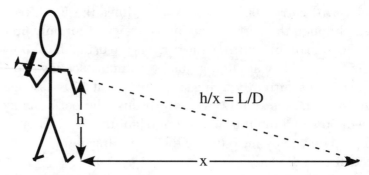

FIGURE 6.3   Determination of *L/D*.

flight equals the weight, so maximum *L/D* simply means minimum drag. In this book we will take the perspective of the pilot and assume that lift is a constant, unless otherwise stated.

Glide

A pilot of a powered airplane must be prepared for the loss of power. Contrary to what many think, an airplane does not fall out of the sky if the engine stops. In fact, an airplane can fly quite a distance without power. Without an engine, the airplane is just a poor-performing glider. Therefore, on loss of power, what should the pilot do?

The pilot may want to maximize the amount of time in the air. This will give more time to search for an emergency landing field, as well as to attempt to restart the engine and to communicate with air-traffic controllers. To achieve maximum time in the air, or endurance, the objective is to minimize the rate of altitude loss. Since the altitude loss is the source of power keeping the airplane flying, the best endurance will occur at the speed for minimum power required (see Figure 1.13). At this speed, the rate of descent is lowest. On the other hand, the speed is slow, and the airplane does not cover much ground as it descends.

What if the pilot is less interested in the time remaining in the air but wants to glide for the longest distance, say to make it to a better place to land? In this case the pilot wants to increase the glide ratio, which is the distance traveled per loss in altitude, and therefore descend at the maximum L/D. In this case maximum L/D is not quite the same as minimum drag because the lift is less than the weight of the airplane. The speed for a glide for maximum range is typically about 20% faster than the speed for maximum endurance (minimum power) and is approximately at minimum drag (Figure 1.16).

Gliders have glide ratios of 25:1 to 60:1. A glide ratio of 25:1 (read "25 to 1") means that for every 1000 ft an airplane descends, it travels 25,000 ft horizontally. This is about 5 miles! A typical airliner has a glide ratio on the order of 16:1, whereas small propeller-driven airplanes have a glide ratio from 10:1 to 15:1. The glide ratio of the Space Shuttle is only 4:1. It has been said that the Space Shuttle "glides like a bathtub."

> The smallest radio-controlled airplane, in 2008, weighed 10g (1/3 oz).

A pilot is trained to know the speed to be flown when the airplane loses power. Part of transitioning to a new airplane is memorizing the new critical speeds associated with that airplane. However, we have seen that power and drag are functions of altitude. Does the pilot need to know critical speeds for every altitude? As you will see in the next section, nature has made life a little easier for the very busy pilot.

OUT OF FUEL

On July 23, 1984, in Ontario, Canada, a Boeing 767 ran out of fuel. An error was made converting from the English system of measure to the metric system. The airplane did not have enough fuel to complete the trip from Montreal to Edmonton. The Air Canada pilot, Robert Pearson, was a glider pilot and was able to bring the 767 down on an abandoned airfield many miles off the flight path.

Let us assume that the Boeing 767 has a glide ratio of 16:1 and was cruising at 32,000 ft (9700 m) when it ran out of fuel. It would be able to glide almost 100 miles (160 km) before landing and would have almost 30,000 mi^2 (80,000 km^2) in which to land.

Indicated Airspeed

We now come to the concept of *indicated airspeed*. In critical maneuvers such as best climb, longest glide, greatest endurance, etc., the pilot must fly at specific airspeeds. The pilot has committed these airspeeds to memory before flying the airplane. However, these speeds change with altitude and air density. So how does a pilot make airspeed corrections for different altitudes and densities? As luck would have it, the airspeed indicator on the instrument panel does it for the pilot, and many of the important airspeeds are marked on it.

The airspeed indicator does not measure the true airspeed of the airplane. It is really measuring the difference in pressure produced by the air striking the end of the *pitot tube* and the static pressure measured at the *static port*, as discussed in Appendix A (see Figure A.10). A pitot tube can be seen on one of the wings of every small airplane. There are several on large jets, mounted near the nose, and they often can be seen as one boards the airplane. Figure 6.4 shows the pitot tube under the nose of a military jet trainer and a pair of pitot tubes on the side of a Boeing 737. The pitot tube is calibrated so that the indicated airspeed and true airspeed are the same under *standard conditions* at sea level. As the airplane flies higher, there is less air striking the pitot tube and thus less total pressure for the same speed. Thus, as the density decreases, the indicated airspeed decreases. The indicated airspeed is lower than the true airspeed at higher altitude. The airplane is actually going faster than indicated. For the pilot to determine the true airspeed in flight, the indicated airspeed must be corrected for air density, which is a function of altitude and temperature.

FIGURE 6.4   Pitot tubes on jet trainer and a Boeing 737.

One might think that having to make these calculations in flight is a nuisance and that the pilot would like the airspeed indicator to read the true airspeed. However, it is the fact that the indicated airspeed is not the true airspeed that makes the pilots life easier. All the critical air-speeds of normal flight are indicated airspeeds. Therefore, even though the speed changes with altitude for such important procedures as climb

> The first Academy Award for Best Picture was awarded to *Wings*.

and glide, the indicated airspeeds of these maneuvers remain the same. When a pilot who is used to landing at a sea-level airport makes a landing at a high-altitude airport, the indicated approach speed is the same. But the ground is going by much faster.

We now have the tools to discuss the performance of the airplane. So we will do just that, starting with takeoff and ending with landing.

Takeoff Performance

The takeoff of an airplane is a fairly simple thing. The airplane accelerates down the runway until it has reached a speed comfortably above the stall speed. The pilot then pulls back on the controls, or *rotates*, and off the airplane goes. The takeoff speed is typically about 20 percent above the stall speed, but it can be as little as 5 percent for some military aircraft. If you were an airplane designer, what other things would you consider in takeoff performance?

The most obvious figure of merit is takeoff distance. If you want to design an airplane that can take off from a short dirt field, you will have to include certain features. If you have unlimited runway, you might design a different airplane. As a general rule, airplanes that have short takeoff distances will fly at lower cruise speeds. Faster airplanes usually need longer runways. Let us examine why this is so.

Since we are concerned with takeoff distance, it is obvious that one can shorten this distance by increasing the engine power or by reducing the takeoff speed. The problem with increasing the size of the engine is that it adds weight and cost. Also, since the parasitic power goes as the speed of the airplane cubed, the increased power will do little for the cruise performance of the airplane. The takeoff speed is decreased by reducing the stall speed, by reducing the wing loading, and by adding high-lift devices such as slats, slots, and larger flaps. These add cost and weight to the wing and also can degrade the cruise performance.

The biggest factor in takeoff performance is the weight of the airplane. The takeoff speed is proportional to the square root of the airplane's weight. A 20 percent increase in weight will cause approximately a 10 percent increase in takeoff speed. The real killer, though, is that the takeoff distance increases with the weight squared. This is a simple consequence of Newton's second law, which states that *acceleration is force divided by the mass*. Thus, if the weight increases, for constant thrust, the acceleration decreases, and it will take a longer distance to reach takeoff speed. Remember that the takeoff speed has increased to exacerbate the problem. Therefore, for example, a 20 percent increase in weight increases the takeoff distance by about 44 percent for a high-powered airplane. But an increase in weight of 20 percent will increase the takeoff distance of a low-powered general-aviation airplane by about 60 percent because of the lower acceleration.

> In March 1945, a C-47 (military version of the DC-3) had its left wing severed off just outboard of the left engine from a midair collision. The pilot managed to make a controlled crash landing with only one wing.

The takeoff distance is also affected by the wind. A headwind that is 15 percent of the takeoff speed will shorten the takeoff distance by about 30 percent, whereas the same-speed tailwind will lengthen the takeoff distance by 33 percent. For a small airplane with a takeoff speed of 70 mi/h (112 km/h), this is only a 10 mi/h (16 km/h) wind.

During the early years of aviation, World War I and earlier, airfields were large square or circular fields. The airplanes could not tolerate crosswinds as they do today, so they would point into the wind for takeoff. When runways were developed, this meant that airplanes could no longer take off in any direction, depending on the local wind. However, careful airport design will place the main runway into the prevailing winds, and sometimes a secondary runway is built between 45 and 90 degrees to it. During World War II, many new airfields were built for training. These airfields had three runways, roughly 60 degrees apart in a triangle. The three runways guarantee a crosswind no greater than 30 degrees.

This is why airplanes always take off into the wind, and aircraft carriers turn into the wind to launch and recover aircraft.

Altitude also contributes to takeoff performance. Recall that it is the indicated airspeed that dictates airplane performance. Thus the takeoff ground speed increases with altitude, although not the indicated airspeed. Here, the difference between a jet and a piston-powered airplane is apparent. The thrust, and thus the acceleration, of a jet engine is less affected by altitude. At an altitude of 6000 ft (1800 m), the takeoff distance of a jet is increased by about 20 percent over the sea-level distance. The altitude affects the acceleration of a nonturbocharged, piston-powered airplane, and thus its takeoff distance is increased by about 40 percent at that altitude.

Small, single-engine airplanes, such as the Cessna 172, can take off in distances less than 1000 ft at sea level. The Cessna 172 has a wing loading of only 13.5 lb/ft^2 (68 N/m^2). The light wing loading contributes to the short takeoff distance.

A Boeing 777 has a high wing loading of over 150 lb/ft^2 (750 N/m^2). This is over 10 times that of the Cessna 172. Its takeoff distance is on the order of 6000 ft (1.8 km). In order to comply with Federal Aviation Administration (FAA) regulations for air transports, which require that the airplane be able to fly safely with one engine inoperable, the thrust available for the Boeing 777 must be double what is necessary for a sustained climb. Thus the thrust available compared with the weight (the *thrust-to-weight ratio*) of a Boeing 777 may be double that of the Cessna 172.

> An average pilot can withstand a force of about 6*g* for a few seconds without blacking out.

A Boeing 777 cannot take off from a 1500-ft (500-m) grass runway. On the other hand, a Cessna 172 cannot cruise at 500 mi/h (800 km/h). Its top speed is only about 125 knots (225 km/h).

Climb

Up to this point, besides glides, only straight-and-level flight has been discussed. You probably have thought about what happens in a climb to increase altitude. A simplistic answer is that you need to generate more lift, the logic being that one increases the angle of attack, the lift goes up, and the airplane climbs. When the pilot first pulls back on the controls to start a climb, that is what happens, but only for a few seconds while the airplane is slowing down. The airplane must slow down because no additional power has been added to create this lift. Then, at this lower speed, there will be less lift than there was initially. In a sustained climb, the lift of the wing is actually less than the weight of the airplane. The forces on the airplane are rotated, except for the weight, as shown in the diagram in Figure 6.5. In this rotated configuration, part of the weight of the airplane is supported directly by the engine. As we will see, the airplane is climbing on the *excess thrust* and *excess power* of the engine. In straight-and-level flight, the thrust and drag are equal. In a climb, a propeller-driven airplane slows down, reducing the drag.

To understand the reduced lift on the wing, let us begin by looking at two extreme situations. First, take the case of straight-and-level flight. The angle of climb is zero, and the lift on the wing is the weight of the airplane. Now consider a very powerful jet fighter that can go straight up in a climb. In this case, the angle of climb is 90 degrees, and the lift on the wing is zero. The engine now supports the entire weight of the airplane. As the fighter goes slowly from straight-and-level flight to a vertical climb, the load on the wing smoothly changes from the weight of the airplane to zero. During this transition, the "lift" produced by the engine goes smoothly from zero to the weight of the airplane.

Now let us consider what happens when a low-powered airplane goes into a climb. Consider a small single-engine propeller-driven air-

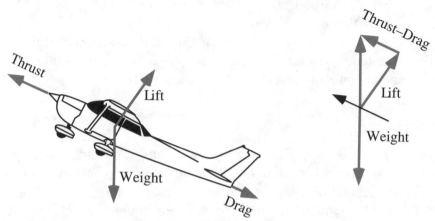

FIGURE 6.5 Forces on an airplane in a climb.

plane flying straight and level at full power. In this case, the wings produce the lift, and the power is just equal to the induced and parasitic power requirements. Now the pilot pulls back a little on the controls, and the airplane starts to climb. Part of the engine's power now goes directly into lifting the airplane. This leaves less power to overcome drag. The airplane slows down, reducing the drag. As the pilot continues to pull back on the controls, the speed is further reduced until the *backside of the power curve* is reached. Eventually, the wing would stall if the angle of attack were increased further.

> Air transportation consumes 8.9 percent of the world's transportation energy budget. Highway transportation consumes 72.4 percent.

There are two climb scenarios of interest to pilots. The first is the fastest climb, or the *best rate of climb*. Airplanes fly more efficiently at higher altitudes, so pilots generally want to climb to their desired cruise altitude as quickly as possible. This is the rate of climb most useful to a pilot. The second scenario is the steepest climb, or the *best angle of climb*. Suppose that you are in a mountain valley and wish to clear the mountaintops. You would want to gain as much altitude in the shortest distance possible and thus fly at the steepest angle. These two scenarios lead to different climb paths and air speeds, which are shown schematically in Figure 6.6.

The best rate of climb for an airplane occurs at maximum excess power. *Excess power* is the difference between the power available and the power required for flight. This is shown in Figure 6.7 for an airplane fitted with either a jet engine or a propeller. The power required is just the power curve discussed in Chapter 1. The arrow

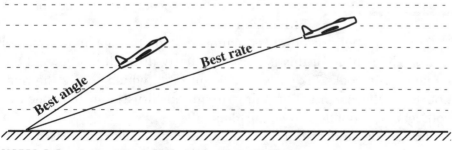

FIGURE 6.6 Best angle of climb and best rate of climb.

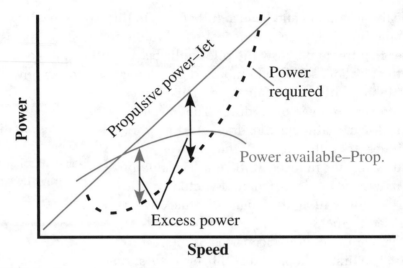

FIGURE 6.7   **Excess power as a function of speed.**

connecting the power curve with the power available for the propeller is the excess power. Excess power is largest at the speed for the best rate of climb. The best rate of climb for a propeller-driven aircraft is at a speed that is very near the minimum drag. One would guess that it would be near the minimum power required, but the variation of the power available with speed causes the best rate of climb to be at a higher speed.

As shown in the figure, the power available from a jet increases with speed. Thus the arrow marking the greatest excess power for the jet is at a considerably higher speed than for the propeller-driven airplane. This is why propeller-driven airplanes slow down to climb, whereas jets maintain their speed or even accelerate a little to climb.

> To be certified, a jet engine must survive the "bird-strike test." The engine must survive the impact of a chicken shot at it with a special cannon. The same cannon is used for testing airplane windshields.

The best angle of climb is achieved with the maximum excess thrust. The reason for this is not particularly difficult but beyond the scope of this book. Take a look at Figure 6.8, which shows the relationship between thrust and drag (required thrust) and the speed of both jet and propeller-driven airplanes.

The available thrust of a propeller-driven airplane decreases with increasing speed. Thus, as shown by the arrow in the figure, the max-

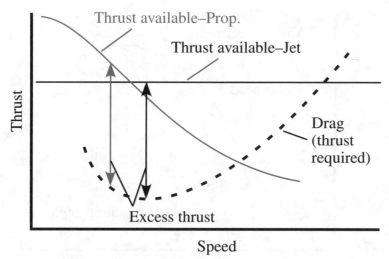

FIGURE 6.8 Excess thrust as a function of speed.

imum excess thrust does not occur at the minimum drag but at a lower speed. In fact, the best angle of climb occurs just above the stall speed of a propeller-driven airplane. Such an airplane, taking off from a short runway with an obstacle such as power lines, will clear them by climbing at about the takeoff speed.

As can be seen in Figure 6.8, the thrust of a jet is approximately constant with speed. Thus the best angle of climb for a jet-powered airplane is achieved at the minimum drag.

For a Cessna 172 at sea level, the best rate of climb is achieved at an indicated airspeed of 84 mi/h (134 km/h). This produces a climb angle of 6 degrees and a rate of climb of 770 ft/min (235 m/min).

The thing to remember is that the best rate of climb is associated with excess power, and the best angle of climb is associated with excess thrust. Next, we will take a look at how high an airplane can climb.

Ceiling

As an airplane climbs, the air becomes less dense. The engine power decreases at the same time that more power is needed to produce lift. What happens is illustrated by Figure 6.9. As the airplane climbs, the minimum speed at which the airplane can fly increases. This is so

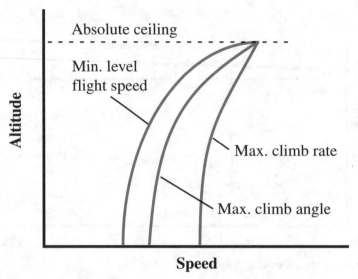

FIGURE 6.9 The minimum flight and climb speeds meet at the absolute ceiling.

because the air becomes less dense, so the minimum speed at which the wing will divert enough air without stalling increases. The speeds for the best rate of climb and the best angle of climb are increasing because the minimums in the power and drag curves are shifting to higher speeds. As the figure shows, at some altitude, the minimum flight speed and the two climb speeds meet. At this point, the airplane cannot fly higher, and full power and thrust are needed to sustain straight-and-level flight. This altitude is the *absolute ceiling*.

As the airplane climbs, its rate of climb decreases. The military defines the *combat ceiling* as the altitude where the best rate of climb drops to 500 ft/min (150 m/min). For some small general-aviation airplanes, this is not much above the runway at sea level. The *service ceiling* of an airplane is defined by the FAA as the altitude at which the airplane's best rate of climb drops to 100 ft/min (30 m/min). This is a useful measure of the performance of an airplane. If one flies out of a high airport, such as the 10,000-ft (3000-m) airport at Leadville, Colorado, an airplane with a 20,000-ft (6000-m) service ceiling certainly is much more desirable than one with a service ceiling of only 14,000 ft (4200 m). The latter airplane will take much more runway to take off and will climb much slower.

Let us look at two high-altitude aircraft that have taken two different approaches for the same mission. The first is the U-2 high-altitude reconnaissance airplane, shown in Figure 2.4. The airplane is an updated 1950s design with a service ceiling over 70,000 ft (21,000 m). This airplane has wings with a very large span and great area. Therefore, the large wings can divert a lot of air very efficiently, and the wing loading is relatively low. The fact that this airplane was slow led to the incident with Gary Powers, when he was shot down over the Soviet Union in 1960. This political catastrophe helped to push the development of a replacement reconnaissance aircraft, the SR-71 Blackbird.

Since the U-2 could not fly fast, it was susceptible to antiaircraft missiles. The SR-71 (shown in Figure 2.5) was developed to overcome this weakness. The SR-71 had to fly at high altitudes and at a very high speed. Its published service ceiling is 80,000 ft (24,000 m), and its maximum speed is Mach 3.2, or 2300 mi/h (3700 km/h). In order to be fast, the SR-71 has a high thrust-to-weight ratio. The compromise for high-speed flight is that the power is high, as is the wing loading. Thus the SR-71 and the U-2, designed for the same mission, made two drastically different choices in their designs.

> The world indoor Free Flight (FF) record as of January 1, 2009, was 60 minutes and 1 second. On June 1, 1997, this was achieved by Steve Brown's *Time Traveler*. Free Flight is a competition using rubber-band-powered airplanes.

Fuel Consumption

When we think of fuel consumption for a car, we think in terms of miles per gallon (or liters per 100 km). These are natural units because cars have odometers, and we measure the amount of fuel when we fill up. On the other hand, these units are not appropriate for an airplane. An airplane is flying in a moving fluid. A small airplane in a strong headwind at a low power setting actually can fly backwards with respect to the ground while measuring a substantial positive airspeed.

Pilots are more concerned with how much fuel is on board and how long they can remain airborne. Recall that in Chapter 1 we saw that induced power is proportional to load squared. The pilot of a commercial airplane wants to fill up with as little fuel as necessary. By the end of the flight, the fuel tanks should contain only the FAA-required reserves, which should be enough to reach an alternate destination, if necessary. The important parameters in determining the necessary fuel are the anticipated ground speed, which gives the time in the air, and the rate that fuel is consumed. The rate of fuel consumption is measured in units of gallons per hour (or liters per hour) for small airplanes and pounds per hour (or kilograms per hour) for large airplanes. Unlike a car, the rate of fuel consumption of an airplane depends only on the power setting and not the ground speed. If the airplane will be flying into a headwind, the ground speed will be lower, the time in flight will be longer, and thus more fuel must be carried.

The *Wright Flyer* was not moved to the Smithsonian until 1948. The Wright brothers were snubbed when the Smithsonian decided to display Samuel Langley's *Aerodrome* as the "first airplane capable of flight." Langley had been the secretary of the Smithsonian.

Now, suppose that you were an engineer and were going to design an airplane from scratch. Clearly, you must consider some other criteria for efficiency. Engineers use a parameter called the *specific fuel consumption*. This is simply the rate of fuel consumption divided by the thrust or power produced, depending on the type of engine. The specific fuel consumption is thus a measure of the engine's efficiency, and the lower the value, the more efficient is the engine. Thus, when choosing an engine, not only must the engineer consider its weight, takeoff distance, cruise speed, and ceiling, but the specific fuel consumption also must be factored in.

Maximum Endurance

If an engine is burning a certain amount of fuel per hour, how long can the airplane stay in the air? Is there a speed at which the airplane can remain in the air longest? Some airplanes are designed for endurance. For example, airplanes built for surveillance may want to loiter over a particular location for a long time. There is interest in using autonomous aircraft that will relay local communications, such as cellular phones, rather than using expensive satellites. These aircraft are mostly concerned about the length of time they can stay in the air. An example is NASA's Pathfinder airplane, shown in Figure 6.10, which was designed to loiter at extreme altitudes and gather atmospheric data. The aircraft is solar powered, so fuel consumption is zero, and

FIGURE 6.10 Pathfinder, a maximum-endurance solar-powered airplane. (Photograph courtesy of NASA.)

the endurance, in principle, is limitless. At night, it runs on batteries charged during the day. Let us see how one would determine how to get maximum endurance out of a fuel-powered airplane.

In maximizing time aloft, speed is not the concern. What is important is the rate at which fuel is burned. The maximum endurance in the air for any fuel-carrying airplane is just the speed of minimum fuel consumption. For a piston/propeller airplane, the power available is almost directly proportional to the engine power. The engine power, and thus the fuel consumption, is just proportional to the required power for flight. Thus the speed at which the pilot should fly a propeller-driven airplane for maximum endurance is at the minimum of the power curve.

Things are different for jet-powered airplanes. As discussed in Chapter 4, the fuel consumption of the engine depends on engine power, not power available. Remember, the power available is just equal to the thrust produced times the speed of the airplane (see Figure 4.3). A jet engine's power available increases with speed, while the thrust remains constant. This means that the jet airplane gets more power for the same fuel flow as the speed increases. Or put another way, for a given power available, the fuel flow can be reduced as the airplane speeds up. This relation to speed means that the minimum fuel flow for a jet is not at minimum required power, as

> In 1997, the Pathfinder set an altitude record for a propeller-driven airplane of 71,490 ft (21.8 km).

it is for a propeller-driven airplane, but rather at minimum required thrust, which is at minimum drag. Thus a jet pilot should fly at the minimum drag speed for maximum endurance. Since minimum drag is a higher speed than minimum power, an airplane with a jet engine will cover more ground flying at maximum endurance than a propeller-driven airplane.

Maximum endurance means maximum time in the air. This is useful for the very few aircraft that have a need to stay in the air for a long time. A more interesting cruise condition, however, is maximum range.

Maximum Range

The best range for an airplane is different from maximum endurance. At maximum endurance, the airplane is traveling quite slowly. Now we want to cover as much ground as possible on the available fuel. As with endurance, the conditions for maximum range are different for a propeller-driven airplane than for a jet.

For a propeller-driven airplane, the power available is roughly independent of speed. For the maximum range, one does not want to find the minimum in power, as for endurance, but the minimum in power divided by speed. Power is equivalent to fuel consumption, and speed is equivalent to distance traveled. Thus power divided by speed is equivalent to gallons per mile, but power divided by speed is just drag. Therefore, for the maximum range, a propeller-driven airplane flies at the speed for minimum drag. This speed is higher than the maximum endurance speed.

Now let us consider a jet. Recall that a jet flies at the minimum drag speed for maximum endurance. This is so because there is a speed term associated with the efficiency of the engine. Thus the argument goes as before, and for a jet, we get the result that for maximum range, a jet-powered airplane flies at the speed for minimum drag divided by speed.

For a propeller-powered airplane, maximum endurance and maximum range are at the speeds for minimum power and minimum drag, respectively. For a jet, maximum endurance and maximum range are at the speeds for minimum drag and minimum drag divided by speed, respectively.

We know what speed to fly for maximum range, but the load of an airplane decreases with time. Therefore, how does an airplane adjust for this reduction in weight to minimize fuel consumption on a flight? For small changes and for small airplanes, changes in weight are adjusted for by changes in the angle of attack. For large jets, though, things are different.

Cruise Climb and Efficiency

Large commercial jets have two constraints on their flight configuration that affect fuel consumption. The first is that to reduce drag, the aircraft must fly at a specific angle of attack so that the fuselage is perfectly aligned with the oncoming air. The second is that the wing is designed to be flown at a certain airspeed for maximum efficiency.

To keep the angle of attack constant throughout the flight, commercial jets prefer to adjust for larger weight corrections by reducing the amount of air the wing diverts. This is accomplished simply by climbing to a higher altitude, where the air is thinner. Thus, as fuel is consumed, the airplane wants to climb to where the air is less dense.

> Eight days after the end of World War I, the world speed record was set at 163.06 mi/h (262.36 km/h).

It is straightforward to understand the change in altitude with change in weight. If a large jet were to become 20 percent lighter owing to the consumption of fuel, the lift must be reduced by that amount. Since it is undesirable to change the angle of attack of the wings or the speed, the vertical velocity of the downwash is a constant. Therefore, the amount of air diverted must be reduced by 20 percent. This means a 20 percent reduction in air density. Thus, as a large jet burns fuel, it wants to fly at an ever-increasing altitude so that its weight divided by the air density remains constant. This is called a *cruise climb*.

As an example of the change in altitude required, a large jet initially flying at 30,000 ft (9100 m) would have to fly at about 35,000 ft (10,600 m) to correct for a 20 percent reduction in weight. Ideally, this change in altitude would be accomplished by a slow climb over the entire flight. Of course, for safety reasons, the Air Traffic Controllers do not allow this. Unfortunately for fuel consumption, jets are only allowed to increase their altitude in 4000-ft (1200-m) steps as they fly.

A pilot can request an altitude change during the flight and thus approach a cruise-climb situation with a *stepped-altitude climb*. On most long flights, pilots will perform a stepped-altitude climb, and it is not uncommon for a pilot to announce such a change in flight. The purpose is to decrease the amount of fuel consumed.

One might ask whether large airplanes fly at the best fuel consumption. The answer is, "Almost." Take a look at Figure 6.11, which shows the range as a function of the airplane's speed for a large jet. Large airplanes usually fly on the high side of the maximum. They would fly at 99 percent of the maximum range because the 1 percent loss in range gives them a several-percent increase in cruise speed. This is considered a good tradeoff. With increasing fuel prices, pilots are slowing down slightly as the exchange between speed and range becomes less favorable.

You should recognize that the range a particular pilot gets from an airplane depends on the pilot's flying habits and how fast he or she wants to get to the destination. Most pilots of general-aviation airplanes fly considerably faster than the speed for maximum range. This

> On one emergency evacuation flight from Ethiopia, an Israeli El al Boeing 747 took off with 1087 refugees and landed with 1090. Three babies were born in flight. A sister ship carried another 920 refugees.

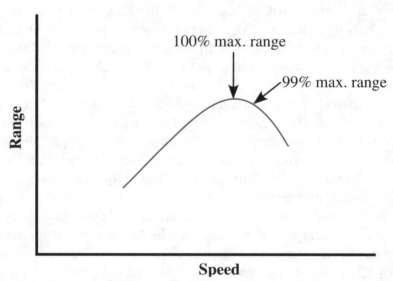

FIGURE 6.11   Range as a function of speed.

is so because a good part of the expense of flying is tallied by the hour and not just the cost of the fuel.

Turns

Now that we have the airplane in flight, let us make a turn. Unfortunately, for commercial and general-aviation airplanes, turns are hardly interesting. If they were, passengers would complain. However, to someone not familiar with a 2*g* turn, this might still qualify as interesting. High-performance turns are primarily the domain of fighter and aerobatic aircraft.

In the discussion of the climb, we saw that the engine, not the wing, is responsible for climbing. But it is the wing that is responsible for a turn. Appendix A shows that the load on the wing (and the pilot) increases when an airplane goes into a bank. In a turn, the *load factor* and stall angle of attack become two critical components in understanding turn performance (see Figures A.6 and A.7). A 2*g* turn, which doubles the load on the wing, is achieved by banking the airplane at an angle of 60 degrees, independent of the speed of the airplane.

The load factor is just like real weight as far as the wing is concerned. We know that the induced power and induced drag vary as the load squared. Since the induced power and induced drag increase as the load squared, they have gone up by a factor of 4 in a 2*g* turn.

Now let us look at an easy (not high-performance) turn with a bank angle of 45 degrees. This turn will put a force on the wings and the passengers only 40 percent larger than in straight-and-level flight. That is, the load factor is 1.4. The induced power and induced drag will be increased to about twice their values in straight-and level flight. The pilot has two choices to compensate for the increased induced power: increase the power available or decrease the parasite power to compensate. The former implies adding more engine power; the later, reducing the speed and increasing the angle of attack. For most general-aviation pilots, a turn is entered at a constant power, the pilot adjusts the angle of attack, and the airplane loses speed.

Table 6.1 shows the performance of an airplane making a 180-degree turn in a 45-degree bank at three different speeds. For a given bank, the radius of the turn increases as the speed squared, whereas the time to make the turn increases as the speed. For all three speeds in the table,

TABLE 6.1 Turn Performance for a 180-Degree Turn with a 45-Degree Bank at Three Different Speeds

Speed, mi/h (km/h)	140 (224)	280 (450)	560 (900)
Radius of turn, mi (km)	0.25 (0.4)	1 (1.6)	3.9 (6.2)
Time to make 180 degrees, s	20	40	80

the forces felt by the passengers will be the same, even though the turns will be quite different. This is shown in Figure 6.12, which illustrates turns at three speeds for an airplane in a 45-degree bank for 20 seconds. The effect of speed on turn performance is quite dramatic.

If the bank in the preceding example were increased to 60 degrees, the load factor would be increased to 2, and the induced power and induced drag would be increased to four times the values in straight-and-level flight. The radius of turn and the time to make the 180-degree turn, though, would be reduced by only 42 percent. Suppose that we increased the bank angle even more. Eventually, there must be some limits because at 90 degrees, the load on the wing and the power required would become infinite.

Suppose that you were flying up a canyon (considered a very bad idea) and you wanted to make a tight turn to get out. What is required for a high-performance turn? How does one make a turn of minimum radius?

The minimum turn radius is limited by three characteristics of an airplane. These are (1) the stall speed with the flaps up, (2) the structural strength, and (3) the power available. Let's look at each of these limits individually.

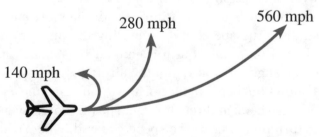

FIGURE 6.12   Turns with 40 degrees of bank for 20 seconds at three different speeds.

Stall-Speed Limit

What is the stall-speed limit? This can be illustrated by considering the extreme of an airplane flying at just above the stall speed (i.e., just below the stall angle of attack). If this airplane tried to make a turn, it could not increase its angle of attack to accommodate the higher load on the wing or it would stall. Therefore, this airplane would be unable to turn and thus would have an infinite turning radius. If the airplane flew a little faster, it could make a very gentle turn with a very large turning radius. The very early airplanes were so underpowered that they could not fly much faster than the stall speed, so they flew into this predicament. They could only make slow turns of large radius. The first flights of many, including the Wright brothers, were made only in a straight line.

Following the logic of the preceding paragraph, let us see what happens when the pilot is in a turn at a speed twice the straight-and-level stall speed. From Chapter 1, we know that at double the speed, the airplane can hold four times the load before it stalls owing to the increase in diverted air and the increase in the vertical velocity of that air. In a bank, this means that the airplane can make a 4g turn (a load factor of 4), which occurs in a 75.5-degree bank. What this shows is that for any given speed, the maximum-performance turn is made at a bank angle, and thus wing loading, just slightly less than what would cause a stall.

For a given airplane, the faster one goes, the higher is the wing loading that can be achieved before stalling in a turn, and thus the tighter the turn. This is illustrated by the curve marked *absolute minimum* in Figure 6.13. The slower the stall speed, the slower the airplane will be going to achieve the same load factor in a turn, and the tighter the turn. Thus the slower the stall speed of the airplane, the tighter the high-performance turn.

> The longest nonstop, unrefueled flight in history was 24,978 mi (40,000 km). This was done by Richard Rutan and Jeana Yeager in the experimental aircraft *Voyager* in 1986.

Structural-Strength Limit

The second characteristic that limits turn performance is the structural strength of the airplane. This sets the limit for the load on the wing and is the curve marked "Structural limit" in Figure 6.13. For airplanes in the *normal category*, which applies to most general-aviation airplanes,

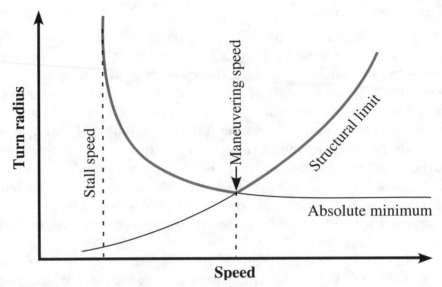

FIGURE 6.13 Turn radius as a function of speed.

the load limit is 3.8*g*. We just saw that a maximum-performance turn at double the straight-and-level stall speed exceeds this limit. Thus the pilot either must reduce the bank angle at this speed or slow down.

If we combine the stall-speed limit and the maximum-load-factor limit, there is a specific condition where the airplane is at both limits. The speed at which this occurs is the *maneuvering speed*. At this speed, the wing will stall just before it exceeds the rated structural limits. Thus, in the normal category, the tightest turn possible would be made at 1.95 (the square root of 3.8) times the stall speed of the airplane and at a bank angle of about 73.5 degrees. That is, the highest-performance turn is made at the maneuvering speed. It would be an uncomfortable turn with high *g* forces.

> A commercial jet can glide for about 100 mi (160 km) if it loses its engines at cruising altitude.

Power-Available Limit

Now let us look at the importance of the power available. As an airplane goes into a steep bank, the load on the wing increases, and the induced power and induced drag increase as the load squared. Thus an airplane in a 4*g* turn experiences an increase in induced power of

MANEUVERING SPEED

The maneuvering speed of an airplane is the maximum speed for an airplane to use during a maneuver or in turbulent air. As discussed earlier, at this speed, the wing will stall at just the maximum rated load factor and thus not exceed it. Since the maximum load factor at stall is proportional to the speed squared, the maneuvering speed of an airplane is just the stall speed times the square root of the maximum load factor of 3.8.

We know that the stall speed decreases with decreasing load. Therefore, the maneuvering speed also decreases with decreasing load. A lighter airplane has a lower maneuvering speed than it would have if it were more heavily loaded. A lighter airplane is flying farther from the critical angle than a fully loaded airplane. Thus it is able to experience a greater acceleration from a gust of air or an abrupt control movement before stalling. The FAA defines the limit for the airplane as 3.8g rather than based on maximum load on the wing. Therefore, at the maneuvering speed, a lighter airplane will stall with a lower wing loading than a heavier one would.

It should be noted that the maneuvering speed is an indicated air speed. Thus the true air speed increases with increasing altitude.

16 times. Therefore, the power available may not be sufficient for the theoretically tightest turn. When this condition is reached, the maximum-performance turn is no longer at the maneuvering speed.

The power available for both jet engines and propeller/piston engines decreases with increasing altitude. At some point, an airplane that is trying to achieve a maximum-performance turn becomes power-limited. Rather than turn at the maneuvering speed, the minimum-radius turn will occur at a somewhat lower speed. Thus, at higher altitudes, the best turn performance decreases.

There is one caveat that allows tighter turns despite the loss of power available. The pilot can choose to "buy" power with altitude. In other words, in a descending turn, the pilot can supplement the engine power with the power it used to climb to altitude in the first place. This is why fighter airplanes engage at high altitudes, but the fight progresses to lower altitudes. The airplanes are using altitude to tighten their turns.

STANDARD-RATE TURNS

The turning rate is usually not a critical design issue for aircraft other than fighter planes and specialized acrobatic airplanes. However, all aircraft must be able to perform a *standard-rate turn*. A standard-rate turn for light airplanes is defined as a 3 degrees/s turn, which completes a 360-degree turn in 2 minutes. This is known as a *2-minute turn*. For heavy airplanes, a standard-rate turn is a 4-minute turn. Instruments, either the *turn and slip indicator* or the *turn coordinator*, have the standard-rate turn clearly marked. This is very useful to pilots who are out of visual contact with the ground and for air-traffic control when appropriate separation of aircraft is desired. The pilot banks the airplane such that the turn and slip indicator points to the standard-rate-turn mark and then uses a watch to time the turn. The pilot can pull out at any desired direction depending on the length of time in the turn.

Landing

What goes up must come down. So another performance parameter is landing distance. Landing distance is easier to understand than takeoff distance. When the airplane approaches its touchdown, it has a certain amount of kinetic energy ($\frac{1}{2}mv^2$). When it comes to a stop, it will have zero kinetic energy. Thus the landing distance will be proportional to the touchdown velocity squared. Therefore, landing performance benefits from a low stall speed.

Typically, the landing distance of an airplane is shorter than the takeoff distance. This is so because the airplane can decelerate with its brakes faster than it can accelerate with its engines. As with most cars, one can stop in a shorter distance than it takes to accelerate to the same speed. There have been many stories of pilots who have landed in a short field but have been unable to take off.

Once on the ground, the airplane's minimum stopping distance will depend primarily on its ability to brake. The braking power is proportional to the weight supported by the wheels. On a hard, dry surface, the decelerating force from the brakes can be as high as 80 percent of the weight on the wheels. Of course, this value is greatly reduced for a slippery surface. Thus, for a maximum-performance stop, the lift must be removed from the wings as quickly as possible to put the weight on the wheels. Therefore, as soon as the airplane touches down, the flaps are raised. Modern jets also employ *spoilers* on the tops of the wings that remove part of the lift and add drag to assist in slowing the airplane down.

One key factor in determining the minimum stopping distance is the ability of the brakes to absorb energy. The brakes of a 500,000-lb (227,000-kg) airplane landing at 170 mi/h (270 km/h) must dissipate 30,000 hp (22,000,000 W) for a half-minute or so! The heat from the brakes can be so great that the wheels will literally melt. This energy that must be dissipated makes the use of thrust reversal important to reduce demands on the brakes.

Commercial airplane tires are filled with pure nitrogen to remove oxygen that can contribute to a fire in case of maximum braking. Also, the heat will cause the tires to expand and potentially blow up. Some airplanes use special bolts that will separate under high temperatures so that the tire pressure will be reduced before the tires can explode. Thus, normally, commercial airplanes do not use maximum braking. However, even after normal landings, airplanes are required to wait for a certain length of time to let the brakes cool. Flights cannot shorten the time before the next departure without artificially cooling the brakes. Cooling times for normal landings are on the order of 30 minutes. Airlines must make sure to schedule airplane turnaround times to accommodate this requirement.

> In a sustained climb, the lift of the wing is actually less than the weight of the airplane.

Light aircraft do not have a problem with the amount of energy that the brakes must dissipate because their landing speed and weight are so low. In addition, these airplanes are usually landed in a high-drag configuration, such as with full flaps, which will slow the airplane down naturally without brakes. Pilots are always happy when they reach a nice rolling speed by the first runway turnoff without using brakes.

As with takeoffs, wind and altitude also affect landings. In fact, they affect landing performance in exactly the same way as they do takeoffs. Earlier we gave an example of a headwind that was 15 percent of the takeoff speed shortening the takeoff distance by 30 percent. A headwind that is 15 percent of the landing speed will shorten the landing distance by the same 30 percent. Likewise, the landing distance at an altitude of 6000 ft (1800 m) will be 20 percent longer than at sea level. This is due to the increase in true airspeed on touchdown and the reduction in drag from the lower air density.

Wrapping It Up

Now that we have discussed performance characteristics, we can look at the performance of some different aircraft. Let us consider a small single-engine airplane, a large transport, and a military fighter. The Cessna 172, the Boeing 777, and the Lockheed-Martin F-22 Raptor will be used as examples here.

The Cessna 172 (see Figure 2.10) has very low wing loading as well as moderate L/D, power-to-weight ratio, and specific fuel consumption (fuel consumption divided by the thrust or power produced). With the low wing loading, the Cessna has a very short takeoff distance compared with the other two example aircraft. However, its cruise speed and range do not compare. Lack of range is due in part to the fact that the Cessna 172 carries less than 10 percent of its weight in fuel. Because of the low wing loading, the Cessna's turning radius is much smaller than that of the other airplanes. However, this is not to imply that the Cessna can outmaneuver the F-22 because the rate of turn is low.

The Boeing 777 (see Figure 2.42) has a high wing loading and high L/D and thrust-to-weight ratio. It also has a low specific fuel consumption. Because of the high wing loading, the Boeing 777 needs long runways found only at the larger, commercial airports. With a maximum L/D 50 percent greater than that of the Cessna and a higher thrust-to-weight ratio, the Boeing 777 has a service ceiling about 2½ times that of the Cessna 172. One design goal of the Boeing 777 was to have a long enough range to service trans-Pacific routes. Therefore, efficient engines are used, and it carries up to 40 percent of its weight in fuel. Neither tight nor fast turns are necessary or possible with this airplane.

A Boeing 747-400 typically takes off at 180 mi/h (290 km/h), cruises at 565 mi/h (910 km/h), and lands at 160 mi/h (260 km/h).

The F-22 (see Figure 2.43) has a high wing loading, moderate L/D, high thrust-to-weight ratio, and moderate specific fuel consumption. The high thrust-to-weight ratio allows for extremely steep climbs and high maneuverability. It also gives a high service ceiling despite the low L/D. Its service ceiling is just a little higher than that of the Boeing

777. The range of the F-22 is low because of the moderate L/D and the desire to trade fuel weight for payload weight. The F-22, like all fighters, requires frequent refueling for long trips.

Chapter 7 will introduce you to the uses of flight testing and wind-tunnel testing. These types of tests are used to determine some of the parameters discussed in this chapter.

Aerodynamic Testing

You have been introduced to many concepts of flight, from lift to performance. One might ask how the designers know that their calculations are correct and, once the airplane is constructed, how the performance is determined. This is the subject of aerodynamic testing, first in a wind tunnel and finally in the air.

Wind-Tunnel Testing

Wind-tunnel tests were performed well before the first airplanes flew. Most of the early wind tunnels were built to perform experiments on fundamental fluid motion. The scientists who built these apparatuses were not interested in flight but instead in the physics of fluids. The Wright brothers can be credited with being among the first to use a wind tunnel to test aerodynamic shapes. Their ingenuity and accomplishments in testing are still a source of amazement after 100 years. Many pages could and have been written about their accomplishments, so we will not go into detail here. An important thing to note is that the Wright brothers recognized the need for useful aerodynamic data and so built a wind tunnel and a device that could measure lift.

Figure 7.1 is a photograph of a wind tunnel that was designed and built in 1918 at McCook Field, Ohio, now at the Dayton Air Museum. It had a 24-blade fan 60 in (152 cm) in diameter that was able to

FIGURE 7.1 Wind tunnel in 1918 at McCook Field, Ohio.

develop airspeeds of up to 453 mi/h (725 km/h). It was used to calibrate airspeed instruments and to test airfoils. The air was drawn in
through the small end of the tunnel and exhausted from the large end.
There was a small glass window (not shown) for observations.

Before delving into the purposes of wind tunnels, let us explore
some of the basic concepts behind the wind tunnel. First, we will discuss the subsonic venturi.

Subsonic Wind Tunnels

A subsonic wind tunnel works like a venturi, shown in Figure 7.2. A
venturi is a good example of Bernoulli's principle, which relates speed
and pressure when no energy is added to flow (see Appendix B). In the
restriction in the tube, the air velocity increases.
The increase in velocity is accompanied by a
reduction in the static pressure, measured perpendicular to the direction of flow. The total
pressure, which is the sum of the static pressure
and dynamic pressure, remains constant and is
measured in the direction of flow. Since the
pressure forces are low, the air density and temperature remain essentially constant. As we will
see later, this is not true in the case of transonic venturis.

The simplest of all low-speed wind tunnels is a venturi. If you have
a household fan and cardboard you can easily build a small wind tun-

> Eleanor Roosevelt shocked military
> officers when she told them she
> wanted to take a ride with an
> African-American pilot from
> Tuskegee.

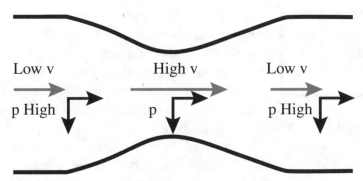

FIGURE 7.2   The venturi.

nel, such as that shown in Figure 7.3. The fan is placed so that it draws air through the wind tunnel. The model is placed in the venturi. The cross section of the wind tunnel does not have to be round. In fact, most wind tunnels have rectangular cross sections.

The change in cross-sectional area of the tunnel from the fan section to the test section is called the *contraction ratio*. If you reduce the cross section of your cardboard wind tunnel to one-fifth the original size, a contraction ratio of 5, the airspeed in the test section will be 5 times the air speed produced by the fan. Thus a fan with a 2-ft (61-cm) diameter that moves air at a speed of 15 mi/h (24 km/h) would produce 70 mi/h (112 km/h) air in a test section about 11 in (27 cm) in diameter. On the other hand, if you want a test section that is 5 ft (1.5 m) in diameter, you would need a fan 11 ft (3.3 m) in diameter. A more practical solution is to increase the fan's air speed and reduce the contraction ratio to achieve a larger test section at the same speed.

There are practical problems in building a venturi wind tunnel. You cannot just contract and expand the walls arbitrarily. This is so

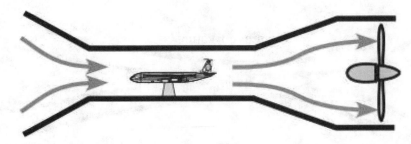

FIGURE 7.3   Wind tunnel.

The Boeing 777 was designed with folding wingtips. No customer ever ordered them.

because the air must constrict smoothly to reduce the effects of the walls. Contracting too quickly will make the walls act as a block to the airflow, as illustrated in Figure 7.4. The pressure buildup limits the effectiveness of the fan.

Expanding too quickly after the venturi also causes problems. The air will not be able to follow the walls, causing the flow to separate from the walls. This also causes a buildup of pressure, which reduces the effectiveness of the tunnel. Figure 7.5 illustrates how a venturi wind tunnel should work.

The wind tunnel just described, where the air passes once though the test section and then exits to the outside world, is called an *open-circuit wind tunnel*. A practical problem with such wind tunnels is that all the energy put into the air is lost and cannot be recycled. This makes the open-circuit wind tunnels inefficient. Therefore, you generally do not see large open-circuit wind tunnels. There is one exception: the 80 × 120 ft wind tunnel at the NASA Ames Research Center, discussed a little later.

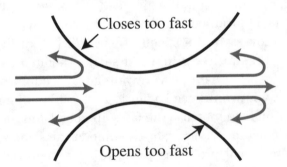

FIGURE 7.4   A venturi that contracts and expands too quickly.

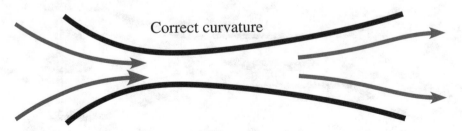

FIGURE 7.5   How a venturi wind tunnel should work.

MEASURING THE AIRSPEED IN THE TEST SECTION

The venturi is an application of the Bernoulli principle. As the air enters the contraction, it speeds up, and the static pressure drops. Since Bernoulli's theorem states that static pressure and airspeed are related (provided that no energy is added to the air), this relationship can be used to determine the airspeed in the test section. Without going through the mathematics, the velocity in the test section is just proportional to the square root of the difference between the static pressure in the test section and the static pressure in the region with the fan. A *manometer* is used to measure the pressure difference. In its simplest form, a manometer, shown in Figure 7.6, is just a bent glass tube filled with liquid connecting the two regions. The difference in pressure is proportional to the difference in height of the liquid, as shown. In the figure, p_1 would be the pressure in the test section, and p_2 would be the ambient pressure. The wind-tunnel operator only has to monitor the difference in height of the liquid and the air density to know the airspeed in the test section. Kerosene is often used as the liquid because it will not evaporate and is safer to handle than mercury.

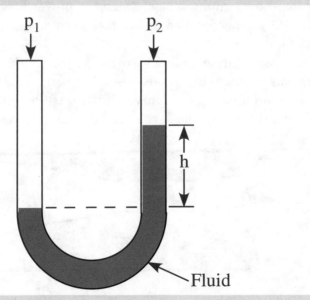

FIGURE 7.6   The manometer.

Closed-Circuit Wind Tunnels

Much of the lost power in an open-circuit wind tunnel can be recovered if you close the airflow circuit, as in Figure 7.7. The closed-circuit wind tunnel is the most common design for larger tunnels. Once the air is accelerated to operating conditions, the fan only needs to add the power that is lost owing to drag on the model and friction on the walls of the wind tunnel.

Be aware that this power loss may be quite significant. Consider a production wind tunnel running many hours a day. The friction and drag on the model can result in considerable heat input to the air and the walls of the wind tunnel. Some sophisticated wind tunnels have cooling vanes to take this heat out and try to maintain a constant temperature. Other wind tunnels just suffer in the heat. Some tunnels can get so hot that changes to the wind-tunnel models must be performed with insulated gloves.

> A wind-tunnel model for a major development program can cost millions of dollars.

Transonic wind tunnels are designed to test aircraft at roughly Mach 0.7 to Mach 0.85. These tunnels require much more power than their low-speed counterparts. Since the power loss owing to friction goes as the airspeed cubed, much more heat is generated and must be removed.

Most wind tunnels have a single return, as shown in Figure 7.7. There was a time when dual-return wind tunnels were popular. The Kirsten wind tunnel at the University of Washington, Seattle, is a dual-return wind tunnel. The layout of this wind tunnel is shown in top

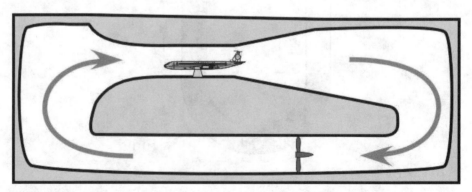

FIGURE 7.7 The closed-circuit wind tunnel.

view in Figure 7.8. The advantage of such an arrangement is that two smaller motors can be used rather than one large motor. There is also an advantage of size with a dual-return wind tunnel. For reasons beyond the scope of interest of this book, a dual-return wind tunnel can support a larger test section while having a smaller footprint. Smaller motors and a smaller footprint result in a lower cost of construction. The disadvantage is that the two channels of air

> **The Wright brothers flew from Kill Devil Hill, a few miles south of Kitty Hawk.**

must meet and become uniform by the time they reach the test section. This causes additional technical difficulties. Figure 7.9 shows a model in the test section of the Kirsten wind tunnel. The test section of this wind tunnel has a cross section that is 8 × 12 ft (2.4 × 3.6 m).

The NASA Ames 40 × 80 wind tunnel is the largest close-circuit wind tunnel in the United States. In the jargon of wind tunnels, a 40 × 80 wind tunnel has a test section that is 40 ft (12 m) high and 80 ft (24 m) wide. This wind tunnel can produce winds up to 350 mi/h (650 km/h). Six fans and six motors, shown in Figure 7.10, drive the wind tunnel. The railing in front of the upper row of fans gives one the feeling of the scale. You also might be able to make out the three people standing in front of fan number 3. The fans are 40 ft (12 m) in diameter with 15 variable-pitch blades. Each motor is rated at 12 MW (18,000 hp).

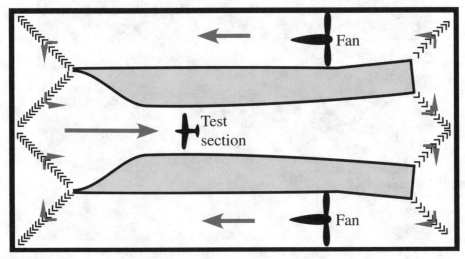

FIGURE 7.8   Dual-return wind tunnel.

FIGURE 7.9   Model in the Kirsten wind tunnel at the University of Washington, Seattle.

FIGURE 7.10   Fans of the NASA Ames 40 x 80 wind tunnel. (Photograph courtesy of NASA.)

FIGURE 7.11 Test section of the NASA Ames 80 x 120 open-circuit wind tunnel. (Photograph courtesy of NASA.)

NASA Ames also has an 80×120 ft (24×36 m) open-circuit wind tunnel. The truly impressive test section of this wind tunnel is shown in Figure 7.11. Here, winds up to 100 mi/h (160 km/h) can be produced. Both the 40×80 and the 80×120 use the same fans, shown in Figure 7.10. The tunnel can be reconfigured by turning vanes to be either an open- or closed-circuit wind tunnel. The original wind tunnel was the 40×80, but in the early 1980s, a plan was put forth to create the 80×120 wind tunnel. This major upgrade allows for large wind-tunnel models to be tested in the 40 × 80 and full-scale aircraft to be tested in the 80 × 120.

One thing to note about both open- and closed-circuit wind tunnels is that the test section is usually kept at atmospheric static pressure. Significant pressures can build up in closed-circuit tunnels, particularly in supersonic wind tunnels. Keeping the test section at ambient pressure allows the use of windows for viewing and photographing the model. Exceptions to this are highly specialized tunnels that have pressurized test sections to increase air density.

Wind-Tunnel Data

What sort of data are collected in a wind tunnel, and how are they used? The most obvious purpose of a wind tunnel is to measure forces and torques, which are twisting forces, on the airplane. Wind tunnels also can be used to measure pressures and airflow patterns on parts of a model. Much of this testing is highly specialized and complicated and so will not be discussed here. However, there are some interesting things to learn about wind-tunnel testing and how it relates to the concepts introduced in this book.

> The hummingbird has a wing area of 2 in² (12 cm²). The albatross has a wing area of 6200 in² (960 cm²).

FORCES

First, we will focus on lift, drag, and torque. A typical wind tunnel uses a *force* and *moment* balance to measure these aerodynamic forces. Figure 7.12 provides an illustration of such a balance. Although the figure shows only one gauge, in reality, a force balance makes measurements of all the forces and torques on the model. In fact, sometimes more than one balance can be used, as can be seen in Figure 7.11. One question is, "Are they accurate?" Several factors enter into the accuracy of wind-tunnel measurements. The most important factor affecting the accuracy of measurements is the effect of the wind-tunnel walls.

The wind tunnel introduces an artificial constraint, namely, walls. The walls have two effects. The first is that they interfere with the amount of air that can be pulled from above the model's wings and block the downwash on the bottom. This latter effect is just like ground effect and is called *wall effect.*

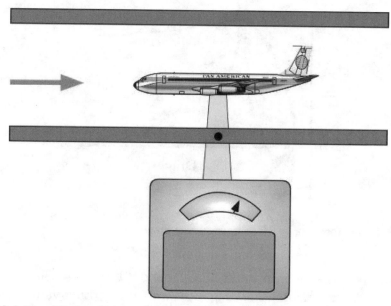

FIGURE 7.12 Wind-tunnel force balance.

Another problem in a low-speed wind tunnel is that air will speed up in a constriction, and the airplane model acts like a constriction in the venturi. In other words, the air accelerates as it moves around the model owing to blockage of the air by the model. For obvious reasons, this is called *blockage effect*.

Years of theoretical work have resulted in methods to correct for wall and blockage effects. Unfortunately, wind-tunnel corrections are not completely reliable, so wind-tunnel results must be backed up with flight tests. Normal wind-tunnel corrections amount to only a few percent. However, the few percent can be extremely important when trying to predict performance of the final airplane. It is not uncommon for an airplane manufacturer to use the same wind tunnel for all tests because the engineers acquire experience in estimating how a particular wind tunnel result will relate to actual flight data.

One of the most important pieces of data collected is called the *drag polar*. This is a plot of lift versus drag, as shown in Figure 7.13. At first glance, these data do not look very interesting, but look again. What makes drag polars interesting is that they can be used to determine the parasite and induced drag as well as the maximum lift before

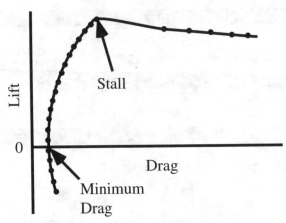

FIGURE 7.13 Test data on lift as a function of drag.

the stall occurs. The minimum-drag value on the graph is just the parasite drag of the model. The drag measured at any other value of lift is the induced drag plus this parasite drag. Since the induced drag and lift are available from these data, the wing efficiency can be determined. Notice that at some point the lift decreases and the drag increases dramatically. This, of course, is the point where the wing stalls and the form drag increases. From this, the poststall characteristics of the wing are determined. A great deal can be learned about an airplane from this single plot.

PRESSURES

A more sophisticated wind-tunnel test also will involve the measurement of pressures on a wind-tunnel model. Tiny holes, called *pressure taps*, are drilled into the model. These taps are connected with tubing to a *pressure transducer*, which is a small electronic device that converts pressure into a voltage to be read by a computer. Some models have over 1000 pressure taps. Thus the pressure measurement system must be able to scan all these pressure taps in a short time.

In 1908, a French woman named Therese Peltier became the first woman pilot.

The pressures can be used to determine many things. They can determine flow separation on a surface, as well as being used to calculate local forces. These pressure measurements also can supply validation data to numerical simulations.

FLOW VISUALIZATION

A third type of data collected in a wind tunnel is visualization data. Unfortunately, in a normal wind-tunnel test, there is very little to see. You cannot see the wind flowing. Since wind-tunnel test sections are closed to prevent data corruption, you cannot even feel the wind. All you hear is lots of noise from the powerful fans. The only way to see what is happening to the air is to have something visible to blow around.

The most common visualization tool is smoke. The problem with smoke is that in order to see details, the wind speed must be very slow. The very low speed can alter the airflow enough to make the smoke results misleading. Another problem with smoke is that in a closed-circuit wind tunnel, the smoke builds up after a while.

> During World War II, the "grasshopper" airplanes, such as the Piper Cub, had the lowest combat losses. In one case, a Piper Cub pilot, jumped by a German fighter, managed to turn, land in a field, and hide the airplane in the time it took the German fighter to turn around.

Another method is to use a mixture of clay and a liquid that evaporates quickly. The clay is very fine and has the consistency of talcum powder. This is painted on the model, and then very quickly the wind is turned on and the model positioned. Once the liquid evaporates, the clay is left in a pattern of the surface flow, as shown in Figure 7.14.

FIGURE 7.14   China clay flow visualization.

Areas of flow separation are easy to spot, as well as patterns of the general flow around the airplane. In the figure, one can see that the last quarter chord of the wing had stalled. This method is very effective but cannot be used if there are pressure taps because the clay will clog the holes.

Tufts of yarn are also useful. Very small tufts with minimal impact on the model can be glued to the model's surface. The tufts follow the direction of the airflow, as seen in Figure 7.15, and can be photographed for later analysis.

A fast-growing technique is the use of a pressure-sensitive paint. This is a paint that actually changes the intensity and shade of a color depending on the local pressure. Thus it is becoming possible to actually see pressures on the surface of a model. This technology is still under development but may become common in the future. There are many other methods of visualization, but to discuss them in detail is beyond the scope of this book.

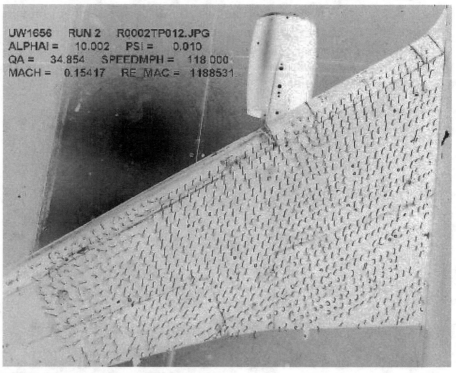

FIGURE 7.15 Minitufts on a model wing.

Tail On and Tail Off

If you visited a wind-tunnel test, there is a good chance that you would see an incomplete airplane. Probably the horizontal stabilizer will not be installed. A normal wind-tunnel test will include many runs without the horizontal stabilizer. There are two reasons for doing this. The first is to be able to provide an estimate of the errors owing to wall effects. The horizontal stabilizer has its own upwash and downwash, which will enter into the measurements. By removing the horizontal stabilizer, the corrections for the wing alone can be determined.

The second reason for removing the horizontal stabilizer is so that the effect of the stabilizer on torques on the airplane in flight can be measured directly. The effect of the horizontal stabilizer on the total torque on the airplane determines how stable the airplane will be, how difficult it will be to fly, and where the payload can be positioned. In analyzing flight characteristics, there are many parameters that cannot be determined theoretically. Thus the wind tunnel provides data that can be used to fill in where theory cannot provide a clear answer.

> The air pressure at 63,000 ft (19,000 m) is so low that water will boil at body temperature.

Supersonic Venturis

In order to understand transonic and supersonic wind tunnels, it is necessary to understand the transonic and supersonic venturi. Here, the compressibility of air cannot be ignored, and both the air density and temperature change significantly. Let us first look at the flow of air that is just below Mach 1 in a tube that decreases in diameter with distance. As the diameter of the tube decreases, the velocity of the air increases. Since the pressure forces are now quite high, as the air compresses, both the density and temperature of the air increase. Because of this compressibility, the velocity and pressures do not change as much as they would if the air were incompressible.

If the tube decreases enough in size, the velocity of the air will reach Mach 1. In a constricting tube, the air does not want to go faster than Mach 1. In fact, in a restricting tube, the speed of the air will always change in the direction of Mach 1. If the air before the restriction is going faster than Mach 1, the dynamic pressure and density will increase at the restriction, slowing the air down until Mach 1 is reached. Once Mach 1 is reached, by either accelerating subsonic air or decelerating supersonic air, the pressure will build up, moving the Mach 1 region forward into a smaller radius in the tube.

The net result is that the Mach 1 region moves to the smallest restriction of the venturi, called the *throat*, as shown in Figure 7.16. What happens after the throat can be somewhat complicated, but in general, it depends on the pressure downstream of the throat. If the pressure is the same as in the tube before the venturi, the air returns to the initial conditions of velocity, pressure, temperature, and density. If the pressure is lower, as in expansion into a large low-pressure volume, the air will expand, causing the velocity to continue to accelerate beyond Mach 1, as shown in the figure. The further expansion and acceleration result in a rapid decrease in air temperature, density, and pressure.

> The radar return on the B-2 bomber and F-117 stealth fighter are roughly the same as that of an eagle.

Supersonic Wind Tunnels

Supersonic wind tunnels operate differently from subsonic and transonic wind tunnels. First, because fans are inefficient at supersonic speeds, they must run subsonically, and the air must make a transition from subsonic to supersonic speeds. Second, supersonic wind tunnels require an enormous amount of power. Supersonic wind tunnels can require so much power that if they are run during periods of peak electricity demand, they can cause a regional brownout. Very few facilities have continuous supersonic wind tunnels for this reason.

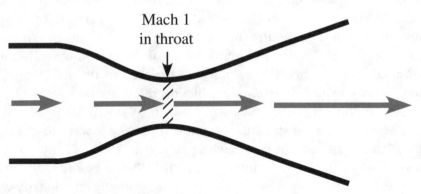

Mach 1
in throat

FIGURE 7.16 Supersonic venturi.

ROCKET MOTORS

It should be noted that this description of a supersonic venturi is the same as the description of a rocket motor. The flow of the compressed gas from the combustion chamber is restricted by the region of Mach 1, which has moved to the throat of the motor. The gas then expands into a region of lower pressure, so it accelerates. Thus the exhaust of a rocket motor is supersonic. Since by Newton's second law the thrust of the rocket is proportional to the velocity of the gas, this is a very desirable situation.

The velocity of the gas is Mach 1 at the throat of all rocket motors. This limits the amount of gas that can be expelled. In designing a motor, the engineer must make the throat small enough so that the gas reaches Mach 1 and thus the exhaust becomes supersonic. But the throat also must be large enough to let enough gas out to produce the desired thrust.

The key to making a supersonic wind tunnel is to employ a supersonic venturi. Figure 7.17 shows a schematic of a closed-circuit supersonic wind tunnel. The fan moves the air in a *subsonic channel*. During startup, the subsonic channel is pressurized, whereas the test section remains at a static pressure of 1 atm. The air accelerates in the first venturi until the speed at the throat becomes Mach 1. As the channel opens up, the air accelerates, producing the supersonic flow in the test section. After the test section, the airflow goes through a second venturi. Here, the speed decreases until it becomes Mach 1 at the throat. Since the air is going into a region of higher pressure, the flow slows down, becoming subsonic again.

The supersonic wind tunnel has an additional source of power loss over subsonic wind tunnels. In addition to the friction on the walls and the drag on the models, now there are losses associated with the inevitable shock

> During an evacuation of a Chinese village in 1942, 68 passengers (and stowaways) were loaded on a DC-3 designed for 30. One of the passengers was Brigadier General James Doolittle, returning from his famous raid on Tokyo.

waves. All these losses mean that a lot of heat is being generated. In order to run continuously, a supersonic wind tunnel must have a large cooler, which is placed in the airflow in the subsonic section.

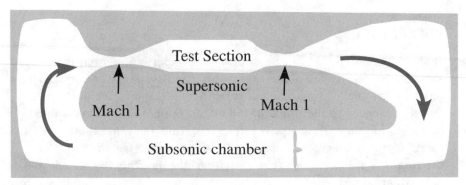

FIGURE 7.17   Supersonic wind tunnel.

The great amount of power required for supersonic wind tunnels means that there are very few continuous wind tunnels, and they are not very large. A 3 × 3 ft (1 × 1 m) test section is considered very large and requires half a million horsepower (375 MW) to operate at Mach 3. However, there are other methods to test supersonic aircraft.

One method is the *blowdown supersonic wind tunnel* depicted in Figure 7.18. A huge tank is filled with high-pressure air and then exhausted through a venturi. This kind of wind tunnel works quite well but will allow only a few minutes of testing. However, a carefully planned test can gather a tremendous amount of data in a very short time. With this technique, the energy required is generated and stored over time. This type of wind tunnel requires very little power but demands quite a long time between tests. The NASA Plumbrook Hypersonic Tunnel Facility can generate airspeeds up to Mach 7. This blowdown facility can accommodate a 5-minute test every 24 hours. The Twenty-Inch Supersonic Wind Tunnel at the Langley Research Center can generate flows with Mach numbers ranging from 1.4 to 5 for 5 to 15 minutes.

> The Boeing 777 carries as much fuel as the Boeing 727 weighs.

Another option, which is more common, is a *vacuum supersonic wind tunnel*, shown schematically in Figure 7.19. Rather than pump a chamber to a high pressure, which is dangerous, the chamber is evacuated, and the airflow is in the other direction through the test section. Thus the upstream reservoir of air is just the atmosphere, and the air is being drawn through the throat and test section into a

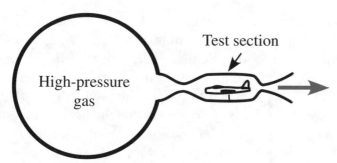

FIGURE 7.18   Blowdown supersonic wind tunnel.

vacuum. Figure 5.2 shows the test of a model of the Space Shuttle in such a wind tunnel.

In all supersonic venturis, the air expands on the high-speed side and thus cools. For continuous supersonic wind tunnels, this is not a concern because all the energy losses cause the air to be hot to start with. For blowdown wind tunnels, the air is often heated before it reaches the venturi so that the test section remains at a reasonable temperature. Vacuum wind tunnels have a problem that the room air is used, and thus it is not practical to preheat the air. Therefore, the test section is very cold. For example, a Mach 3 test section would be −274°F (−170°C) if the air supply were at room temperature.

Hypersonic Testing

With the incredible power required for supersonic wind tunnels, how can anyone expect to create hypersonic flow conditions, typically above Mach 5, in a test environment? The only effective way to do this

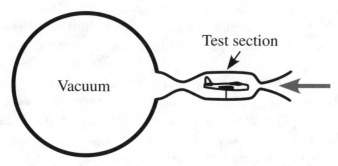

FIGURE 7.19   Vacuum supersonic wind tunnel.

with a stationary model is with the blowdown method, lots of pre-heating of the air, and a very small test section. The key word in that last sentence is *stationary*. Some hypersonic facilities actually use a combustion gun, where gases combust in the breach to propel the model. The problem with this technique is that the desired measurements must be made on a nonstationary model—one that is moving very fast.

> In April 1949, the *Sunkist Lady* stayed aloft for 1 minute over 6 weeks (1008 h, 1 min). The airplane was refueled by flying low over a Jeep that passed up fuel cans.

However, there is another trick an engineer can employ. Hypersonic flight implies that the Mach number is typically greater than Mach 5. Up to this point, we implicitly assumed that to achieve hypersonic speeds, we have to increase the speed of air in the test section or of the model. What if we were to decrease the speed of sound instead? Sound speed differs for different gases. The speed of sound decreases as the weight of the gas molecules increases. Thus, instead of using air for our working gas, we could look for a heavier gas, such as carbon dioxide, although this will only decrease the sound speed by 14 percent. The advantage of using an alternate gas is that the true speeds can be kept reasonable while the Mach number can be fairly high. However, using an alternate gas will not yield information on chemical reactions characteristic of hypersonic flight.

Flight Testing

This section will explain flight-testing techniques used to verify airplane performance as related to the concepts described previously. First, it must be understood that flight testing means two very different things to commercial airplane manufacturers and to military airplane manufacturers. To commercial airplane manufacturers, flight testing focuses on meeting Federal Aviation Administration (FAA) requirements. Frequently, an FAA representative will be on the airplane to monitor the test results. These tests are performed to verify compliance with specific regulations. Because of the expenses, flight tests of commercial transports are rarely performed unless a regulation is involved.

To the military, flight tests usually mean compliance with military specifications. This typically means verifying performance. Because military aircraft fly close to the edge of their operating capabilities, flight testing is used to probe the limits of the airplane. Below we will discuss some of the measurements performed in a flight test.

Flight Instrument Calibration

One of the first steps in flight testing an aircraft is to ensure the altimeter and the airspeed indicator are giving proper information. In most airplanes, the altimeter and airspeed indicator use pressure to determine altitude or speed. We will start with a little explanation of how they work and then discuss what is tested early in a flight-test program.

An altimeter is nothing more than a simple pressure gauge or barometer. The static pressure is measured at the static port, as discussed in Appendix A. The static port is placed somewhere on the surface of the airplane. Because of the airflow around the airplane, the surfaces see a variety of static pressures that are different from the true ambient pressure. However, just because the air is flowing faster somewhere does not mean that the static pressure is lower. This statement is discussed in Appendix B. Since any place on the airplane is going to see at least some small pressure difference from the ambient static pressure, it is important to calibrate the altimeter.

The goal of the altimeter is to measure the pressure of the atmosphere surrounding the airplane with as little airplane interference as possible. Typically, static ports are placed on the fuselage away from the wing, where pressure changes are smallest. Because the static pressure changes with position on an airplane, the error associated with placement of the static port is called *position error*. A flight test must be performed to determine the amount of this position error. To do this, the pressure must be measured free from aircraft interference. One method used by commercial airplane manufacturers is to use a *drag cone*. This is a probe far to the rear of the airplane, as shown in Figure 7.20. In the figure, the drag cone is deployed from the tip of the vertical stabilizer. The drag cone is

> Airplanes need a lot of fuel. A car is about 5 percent fuel by weight; a city bus, 2 percent, a passenger train, 1.1 percent; and a freight train, 0.4 percent. A Boeing 747 is 42 percent fuel by weight.

deployed and then retracted with a winch in the tail of the airplane. Figure 7.21 shows another airplane with the drag cone stowed in the tail of the fuselage.

FIGURE 7.20 Drag cone on a Boeing 737. (Used with permission of Boeing Management Company.)

FIGURE 7.21 Drag cone stowed in the tail of the fuselage.

FIGURE 7.22   Airspeed calibration probe on D-557.II. (Photograph courtesy of NASA.)

On many military tests, flight testing for position error is measured on a probe mounted on a long "spike" on the nose of the airplane, as shown in Figure 7.22. The difference between the true air pressure and that measured by the static port can be translated into an altimeter error, which must be published for the airplane.

THE STANDARD DAY

The pressure variation with altitude is not exactly the same every day, so a *standard day* has been defined and used internationally. A standard day is 59°F and 29.92 inHg (15°C and 1013 mbar) at sea level. The pilot adjusts the altimeter for the true barometric pressure so that correct altitude is indicated. When calling in for permission to land at an airport, among the first things given to the pilot are the barometric pressure and the wind's speed and direction. Below 18,000 ft, all altimeters are adjusted to local atmospheric conditions so that two airplanes in the same airspace will both read correctly. Above 18,000 ft, all altimeters are set for the barometric pressure of a standard day.

Power Required

Once the altimeter and airspeed indicator are calibrated, they can be used, along with engine power calculations for piston-driven airplanes or thrust for jets, to determine the power and drag curves for straight-and-level flight. In principle, one can get both curves and determine the induced and parasite powers and drags by measuring the power at only two speeds.

The preceding measurements were made at one weight and altitude. However, we know that both weight and air density affect the

POWER-REQUIRED DATA

Although a little beyond the scope of this book, if one plots power times speed as a function of speed to the fourth power, the result is a straight line. This is illustrated in Figure 7.23, taken from the flight manual for a Cessna 172. Although values for four points were used, in principle, the same results could be obtained from only two points. From this plot, one can produce a plot of power as a function of speed (the power curve) as well as a plot of drag (power divided by speed) as a function of speed (the drag curve). Knowing how induced power and drag go as speed, one is also able to separate out the different components of power and drag. This is a lot of information from two simple measurements.

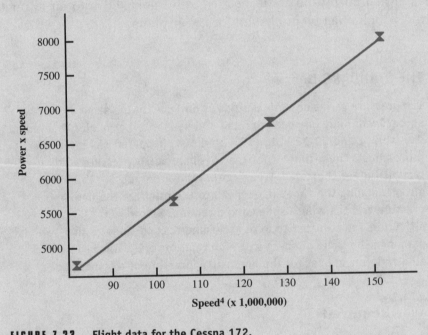

FIGURE 7.23   Flight data for the Cessna 172.

power required for straight-and-level flight. To perform the test described earlier for different altitudes and loads would be expensive and time-consuming. Engineers have developed the *equivalent weight system* to avoid these additional tests. The air density at the test altitude is determined from pressure and temperature measurements and related to sea-level air density. Similarly, the weight of the airplane during the test is measured and related to a standard weight, usually either the maximum gross weight or the empty weight. Known relationships already discussed for power requirements as a function of weight and air density allow the data to be related to any altitude and weight regardless of actual altitude and weight flown on the flight test.

Takeoff and Landing

Flight testing of takeoff performance is one of the more extensive tests required for commercial airplanes. Certain aspects, such as friction with the ground, depend on runway conditions, such as the presence of ice or water. The designer made assumptions about friction, and these have to be verified in flight tests. Once a few conditions are verified, calculations are used to fill in the rest of the operating procedures.

Takeoff techniques also have to be established. For example, at what speed should a pilot begin rotation? At what speed should the airplane lift off? If the speed is too low, the airplane might be in ground effect on the backside of the power curve and unable to climb.

Two tests are particularly exciting. They are the Vmu (velocity-minimum-unstuck) test and the maximum-braking test. Vmu is the minimum speed at which the airplane can leave the ground (that is, unstuck). This occurs at approximately the stall angle of attack. In order to achieve this goal, part of the airplane tail actually may drag on the ground. When aluminum hits concrete, sparks tend to fly. It can look like the airplane is on fire, as shown in Figure 7.24. Usually, little damage is done to the airplane. For flight testing, a small tail skid is placed on the airplane to prevent damage.

> Models for the first Boeing 747 completed more than 15,000 hours of wind-tunnel testing.

The maximum-braking test demonstrates that the airplane can abort takeoff and stop on the runway without risking the passengers. The airplane is accelerated to the takeoff speed and then brought to a stop with only the brakes. What makes this test exciting is that all the kinetic energy of the airplane is transferred to the brakes. Thus the brakes get extremely hot, as discussed in Chapter 6. The test requires that the airplane be able to remain on the runway without help for a certain period of time. You have to understand that the tires are probably melting and exploding. The heat from the brakes may be radiating

FIGURE 7.24 Vmu test with a tail skid. (Used with permission of Boeing Management Company.)

to the underside of the fuel tanks. The test ensures that if a maximum-braking abort is required, the airplane can survive until fire trucks can arrive on the scene to cool down the brakes.

Takeoff and landing tests must be performed in a variety of configurations. These are demonstrating the "What happens if?" scenarios. Tests with one engine out, flaps in various positions, different gross weights, and at different atmospheric conditions must be performed.

As computer models get more and more accurate, some flight testing is being replaced by careful calculations. Thus, rather than having to test every possible condition, which is very expensive, a few key conditions can be tested in flight and used to validate the calculations. The rest of the conditions then can be determined by analysis. The predictive abilities are so good today that there are rarely surprises in performance when the airplane goes for its flight test.

Climbing and Turning

In Chapter 6 it was shown that climbs and turns require power and thrust. Thus, climbing and turning tests are nothing more than determining the available power. The difference between the power available and the required power is the excess power. The bottom line is that the power and thrust available are the only information required to determine climb and turn performance.

To measure the available power, the procedure is to push the throttle to a specific setting and measure the acceleration. The acceleration times the mass of the airplane is the net thrust. Since the required power and thrust can be calculated from a previous flight test, the available thrust and available power can be determined.

Flight-Test Accidents

Flight-test accidents do not leave much to talk about. Today, there is rarely a crash of a commercial airplane in flight testing. With military aircraft, it happens on rare occasions, but nothing like during the rapid development period of the 1940s to the 1960s. Today, even military aircraft performance is so well predicted that there are few surprises. In general, though, the riskier business of military flying will lead to more flight-test accidents. The bottom line is that flight testing occurs all the time with very few accidents.

> The Wright brothers were not the first to use a wind tunnel. However, they were the first to use it for the purpose of understanding flight.

Wrapping It Up

Testing is an integral part of any airplane development program. It is an expensive part too. The days are long gone when testing was used in place of detailed analysis to develop an airplane design. Testing is now used mostly for refinement and validation. Next, we will discuss some very different aircraft—helicopters and autogyros.

Helicopters and Autogyros

Helicopters are very complicated aircraft. When compared with fixed-wing airplanes, they are slow, inefficient, unstable, and difficult to fly. Where they excel is in vertical takeoff and landing and the ability to fly low and slow and to stand still (to *hover*) in the air. This makes them an "angel of mercy" to those in distress, as well as a formidable weapons platform.

In order to keep within the scope of this book, the discussion will be limited to single-rotor helicopters. A classic example is the Bell-47, shown in Figure 8.1, which was made famous in the Korean War. The large blades above the cockpit form the *rotor*, which provides the lift and propulsion of a helicopter. The smaller blades at the end of the fuselage form the *tail rotor*. The tail rotor's primary function is to cancel the torque caused by the engine powering the rotor and to keep the fuselage pointed in the desired direction.

> The Bell-47 became the first helicopter to be licensed for civil operation in January 1947.

To give you an idea of the range of sizes involved in single-rotor helicopters, the Bell-47G has a maximum takeoff weight of 2347 lb (1067 kg) and is powered by a single 200-hp engine, and the Mil Mi-10 (Figure 8.2), produced in the 1960s in the Soviet Union, is also a single-rotor helicopter with a gross weight of 95,100 lb (43,245 kg) and is powered by two 5500-hp engines. An

FIGURE 8.1 Classic Bell-47. (Photograph by Colin Hunter.)

FIGURE 8.2 Mil Mi-10 produced in the 1960s in the Soviet Union. (Photograph by Ken Elliott.)

extreme example of a two-rotor helicopter is the Mil Mi-12, which is powered by four 5500-hp engines.

Before we go on, we would like to dispel one commonly held belief about helicopters. That is that they fall out of the sky on loss of power. It is a requirement of any practical aircraft that, on loss of power, it be able to maintain control and return to the ground at a speed reasonable for landing. The helicopter is no exception. The way the helicopter manages this is by *autorotation*. This is a procedure by which the helicopter maintains lift by exchanging potential energy (altitude) for power to the rotor. This will be discussed in considerable detail later. We will end the chapter with a discussion of the autogyro, which is an aircraft that uses an unpowered rotor to produce lift and a propeller to produce propulsion. First, however, we must discuss the mechanical complexities of the helicopter.

> **The world altitude record for a helicopter is 40,820 ft, set on June 12, 1972, by an Aerospatiale SA-315 Lama, piloted by Jean Boulet.**

The Rotor

The blades of the rotor are merely rotating wings that produce lift in the familiar way—by diverting air down. Like a wing, the lift on a rotor blade is proportional to the effective angle of attack. Unlike a wing, the speed of the rotor increases with distance from the root. In Chapter 1 we saw that the lift on a wing is proportional to amount of air diverted times the vertical velocity of the air. Since both the amount of air and the vertical velocity of the air are proportional to the speed of the wing, lift is proportional to the speed squared. Thus the lift of the rotor increases rapidly with distance from the central hub. The colored line in Figure 8.3 shows the lift produced by a rotor, with a constant angle of attack in hover. Clearly, most of the lift is produced near the tips of the blade. To improve this, many rotor blades are constructed with a twist such that the angle of attack decreases with distance from the hub. This twist gives a more uniform lift distribution, as shown by the black line in the figure. Because of the flexibility of the rotor blades and the relatively high loading away from the hub, the rotors form a cone as they spin.

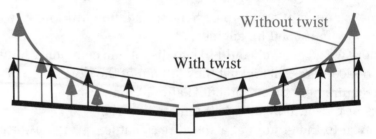

FIGURE 8.3 Lift distribution of rotor blades with and without a twist.

Tip-Path Plane

The rotor *tip-path plane* is a disk (often referred to as the *rotor disk*) formed by the rotor tips, as illustrated in Figure 8.4. The force produced by the rotors is perpendicular to this plane. When the tip-path plane is tilted, as illustrated in Figure 8.5, the force is also tilted. This force can be separated into two components. The vertical component is the lift, used to overcome the weight of the helicopter and to climb in altitude. The horizontal component of force is called the *thrust* and is used for forward motion, turns, or to slow the helicopter down.

The rotor blades revolve in a cone, as discussed earlier. Unlike wings, the rotor blades must be able to swing (via hinges) on three axes besides the rotation about the *central hub*. The three hinges are shown in Figure 8.6, which depicts a rotor blade as viewed from above. In forward flight, the blades swing up and down (*flapping*), change in pitch (*feathering*), and swing forward and back from the radial position (*lead lag*). In forward flight, the tips of the rotor trace an ellipse around

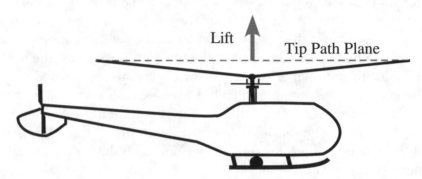

FIGURE 8.4 Rotor tip-path plane.

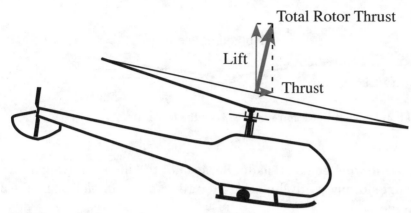

FIGURE 8.5 Tilted tip-path plane and thrust.

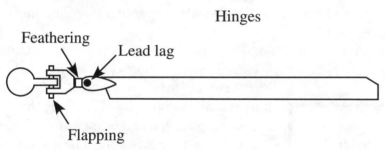

FIGURE 8.6 Three rotor hinges.

what their position would be in hover. These motions will be discussed in detail below.

Flapping

Since the blades are attached to the central hub by the flapping hinge, a restoring force (other than gravity) is needed to keep the lift on the blade from swinging up. To the rescue comes centrifugal force. Centrifugal force is the force that one feels when swinging a weight on a rope. As shown in Figure 8.7, *centrifugal force* tries to keep the blade flat. This force is substantial. The load on each blade root of a two- to four-passenger helicopter can be as high as 12 tons, whereas large helicopters can see loads as

> If the engine stops, the helicopter's rotor continues to spin, allowing the machine to land slowly, generally without crashing to the ground.

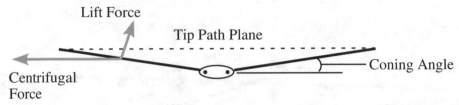

FIGURE 8.7   Centrifugal force tries to keep the blade flat.

high as 40 tons on each blade. Because of the high centrifugal forces, the angle formed with the horizontal (the *coning angle*) is small—2 to 5 degrees.

As will be discussed later, in forward flight, the lift on the blade varies as it makes its revolution. This causes the coning angle to vary, making the blade rise and fall, called *flapping*. Flapping has several consequences.

The first is that as the lift increases, so does the coning angle. The increase in coning angle causes the direction of lift to tilt in, as shown by the colored arrow in Figure 8.8. This reduces the vertical component of lift felt by the aircraft. A second consequence of flapping is shown in Figure 8.9, a view of a rotor blade seen end on. A nonflapping blade will see a certain apparent oncoming wind and have one angle of attack. A blade that is rising (or flapping up) will have a smaller angle of attack and thus a reduced lift. The converse is also true. A dropping blade will have a higher angle of attack and an increase in lift. As will be seen in the discussion at the end of this chapter, this flapping hinge was the key breakthrough in the development of auto-gyros and also made helicopters possible.

> For a week every summer, the busiest airport in the world is Whitman Field in Oshkosh, WI. This occurs during the Experimental Aircraft Association (EAA) annual fly-in.

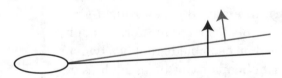

FIGURE 8.8   An increase in coning angle reduces the vertical component of lift.

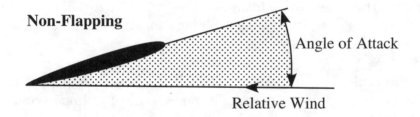

Non-Flapping

Angle of Attack

Relative Wind

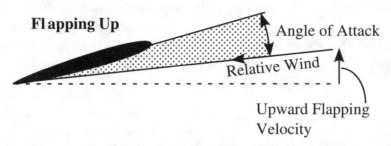

Fl apping Up

Angle of Attack

Relative Wind

Upward Flapping
Velocity

FIGURE 8.9 A rising blade has a smaller angle of attack and less lift.

Feathering

If the rotor blade had a constant angle of attack throughout its rotation, the net lift would be straight up, as shown in Figure 8.4. The helicopter would be able to climb but would have no means of forward flight. Changing the lift of the rotor blades with position around the rotor's path provides forward propulsion or, in general, propulsion in any desired direction. In nonvertical flight, the pitch and thus the angle of attack of the rotor changes with position as it rotates around the axis. This change in pitch (shown in Figure 8.10) is called *feathering*. The axis about which the angle of attack is rotated is the *feather-*

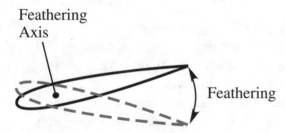

Feathering
Axis

Feathering

FIGURE 8.10 Change in pitch about the feathering axis.

ing axis. The difference in lift around the rotor disk causes the lift vector to tilt, as shown in Figure 8.5, giving a forward-thrust component.

Lead Lag

The third axis is the *lead-lag axis.* As the lift varies around the disk, so does the induced drag. This drag is opposite the direction of rotation, as shown in Figure 8.11. If the lift were constant around the revolution, as it is in hover in still air, the induced drag also would be constant. However, since the induced drag varies around the disk, the variation in drag force causes the blade to lead the constant-drag position at times and at other times to lag. The lead-lag hinge is used to eliminate the bending torque that otherwise would occur at the blade root.

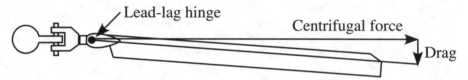

FIGURE 8.11   **Drag is opposite the direction of rotation.**

Rotor Controls

The controls of a fixed-wing airplane consist of a yoke (or stick) for the aileron and elevator control and two rudder pedals. The helicopter is controlled with a *cyclic*, which behaves similar to the stick in an airplane; two pedals for the tail rotor, which behave similar to rudder pedals; and the *collective*, which adjusts the pitch of all the rotor blades equally. We will start with the *swashplate*, which is how commands are communicated from the cockpit to the rotor.

Swashplate

The swashplate, shown in Figure 8.12a, consists of two plates, one that rotates with the rotor and one that is fixed with respect to the helicopter. A bearing joins the two plates. The upper plate is attached to one edge of each rotor blade by a *pitch change arm*. The lower plate is connected to the controls by hydraulic servos that allow the swashplate to be raised and lowered as well as tilted. As the swashplate is

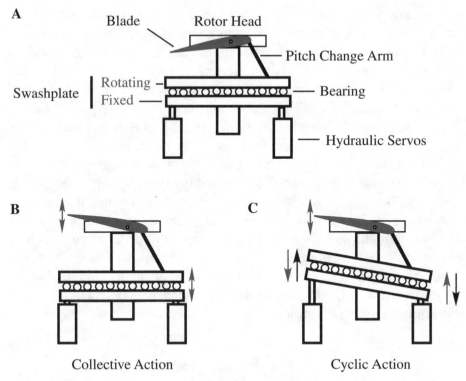

FIGURE 8.12   Swashplate operated by collective and cyclic.

raised and lowered, as shown in Figure 8.12b, the pitch of all the blades decreases and increases.

The collective, a handle to the left of the pilot's seat, is used to raise and lower the swashplate. This changes the pitch of all the rotor blades. The cyclic, a control that looks much like the stick in an airplane, tilts the swashplate, as shown in Figure 8.12c. Thus, as the blade rotates, the pitch changes with position. A tilted swashplate tilts the tip-path plane. In order to gain forward motion, the tip-path plane is simply tilted forward with the cyclic.

> H. Ross Perot, Jr. (son of Ross Perot) and Jay Coburn piloted the first around-the-world flight of a helicopter in 1982.

Gyroscopic Procession

Gyroscopic procession is a phenomenon such that when a force is applied to a rotating object, the effect is seen 90 degrees later. Demonstrations of gyroscopes are very common in physics classes and science museums. A fairly common demonstration is to hold a spinning bicycle wheel by two handles on either end of its axle. Assume that the wheel is spinning counterclockwise when viewed from above and that the handles are vertical. If one tries to tilt the wheel away from oneself, the wheel will in fact tilt to the left. This is an example of gyroscopic procession.

The helicopter's rotor is very much like the spinning bicycle wheel or, more accurately, like a gyroscope. Because of gyroscopic procession, the cyclic control must affect the rotor disk 90 degrees before the desired action. To help understand what is meant by this, look at Figure 8.13. The rotor is revolving counterclockwise, as seen from above. The intention is to tilt the disk down at point *B* and up at point *D*. To do this, though, the downward force must be applied at point *A* and the upward force at point *C*. Thus the angle of attack of the rotor blade is greatest at point *C*, yet the rotor tips are highest at point *D*. This is complicated and certainly confusing. However, it is consistent with the complexity of helicopters.

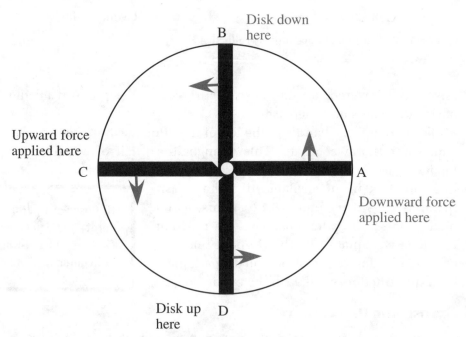

FIGURE 8.13 Because of gyroscopic procession, the cyclic control must affect the rotor disk 90 degrees before the desired action.

Tail Rotor

The primary purpose of the tail rotor is to compensate for the torque from the main rotor and thus to stop the fuselage from turning in a direction opposite the main rotor. This rotational torque is created because the engine is driving the rotor from the fuselage.

The tail rotor is mechanically linked to the main rotor, and thus its rotation rate cannot be adjusted separately from the main rotor. The thrust of the tail rotor is adjusted by varying the pitch with foot pedals. The pitch can be adjusted to give both positive and negative pitch and typically has a range of travel of 40 degrees.

There is one undesirable result from the thrust of the tail rotor. While canceling the main rotor torque, it is also pushing the helicopter to the right. This is called *translating tendency*. Because of this, the main rotor is often tilted

> The rotor in vertical autorotation is about as effective as a parachute of the same diameter as the rotor.

slightly to the left in hover. This causes the fuselage to tilt slightly, with the left skid low. This is why one often sees single-rotor helicopters land on one skid (typically the left) first. Another drawback of the tail rotor is that it consumes power but does not provide lift. The larger the helicopter, the greater is the power load of the tail rotor.

There are cases where single-rotor helicopters do not need a tail rotor. Several small helicopters have been built with small ramjets on the ends of the blades for power. Since the rotors themselves provide the power, there is no torque on the fuselage. Likewise, a helicopter descending in autorotation needs only a very small amount of input from the tail rotor to overcome the torque caused by the friction in the bearings.

Helicopter Flight

Having discussed much of the complications and technology of the helicopter, it is now time to put it all together and discuss forward flight. It should be clear what the main rotor is doing in a hover in calm air. The rotor blades are at a constant angle of attack, the air is diverted down, and the lift is straight up, overcoming gravity. Also, the lift is uniform across the rotor disk. Things change when the helicopter starts to move forward.

Lift Asymmetry

The relative airspeed distribution, as seen by the rotor blades, in a hover is shown in Figure 8.14. Here, the speeds are in knots and depict the relative airspeeds seen for various radii. As the helicopter gains forward airspeed, that speed must be added to the speed created by the moving blades. Figure 8.15 shows the airspeeds for a forward speed of the same helicopter flying at a speed of 150 mi/h. Now the advancing blade sees a much higher airspeed than the retreating blade. The lift on a blade (or a wing) is proportional to the speed squared for a fixed angle of attack. Thus, if the blades had a constant angle of attack, the right side of the rotor disk would have a much greater lift than the left, and the helicopter would tip over. This in fact was one of the early problems in the development of helicopters and autogyros. The lift is kept constant by feathering the blades as they make their transit around the disk. Blade flapping also improves the situation by increasing the lift of the retreating blades and decreasing the lift of the advancing blades.

In forward flight, there is a region of the disk where the air is flowing backwards across the blade. That is, the air is moving from the

> Insects produce lift just like airplanes—by blowing air down.

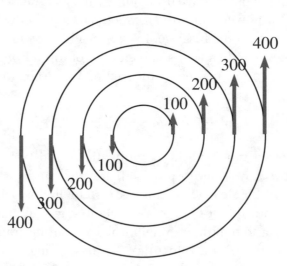

FIGURE 8.14   Relative airspeeds in hover.

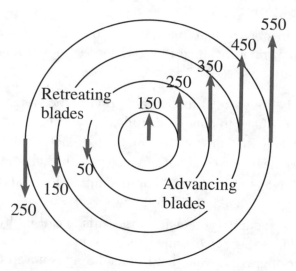

FIGURE 8.15   Relative airspeeds for a forward speed of 150 mi/h.

trailing edge to the leading edge of the rotor blade. This is the region of reverse flow, shown in Figure 8.16, and it changes in size and distance from the hub with changes in forward speed. The region of reverse flow does have one small advantage: It puts a force on the rotor in the direction of rotation and helps it to spin.

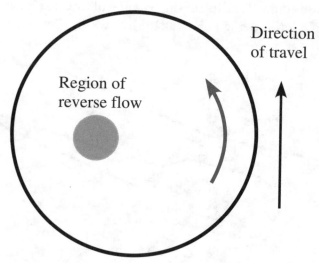

FIGURE 8.16   Region of reverse flow.

Balancing of Forces

A helicopter in steady-state flight, that is, one that is not accelerating or decelerating, must balance the four forces shown in Figure 8.17. These forces are

1. *Lift.* This is the vertical force that supports the weight of the helicopter and is used to change altitude. Lift acts on the rotor hub.
2. *Weight.* This is the vertical force owing to gravity. The weight of the helicopter acts on the center of gravity (cg) of the helicopter's fuselage.
3. *Thrust.* This is the horizontal component of the total thrust, and it acts on the rotor head of the helicopter.
4. *Drag.* More accurately parasitic drag, this is caused by the motion of the fuselage through the air. This is a horizontal force that acts on the center of gravity. Note that induced drag is not in the direction of forward motion in a helicopter.

There are three requirements for the forces in steady-state flight:

1. Thrust must equal drag.
2. Lift must equal weight.
3. The resultants of lift and thrust and of weight and drag must line up, as shown by the "Resultants" in Figure 8.17.

For a better understanding of the balance of forces, look at the three examples in Figure 8.18. Figure 8.18a shows the forces acting on a

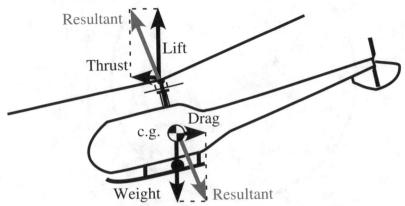

FIGURE 8.17   Balanced forces for a helicopter in steady-state flight.

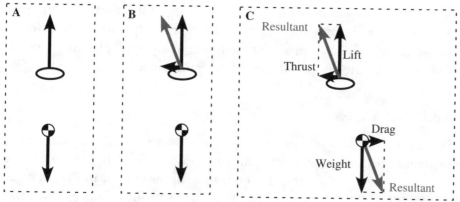

FIGURE 8.18 Balance of forces: (a) hover; (b) just starting to accelerate; (c) steady-state flight.

helicopter hovering in still air. The lift is just equal to the weight, and there are no thrust and drag forces. In Figure 8.18b, the helicopter has just started to accelerate. Here, there is thrust but essentially no drag, and the resultants do not line up. At this point, the fuselage is beginning to swing back, and the speed is increasing. When the speed has increased and the drag equals the thrust, steady-state flight has been reached, as shown in Figure 8.128c. This movement of the center of gravity is what causes the helicopter to tilt nose down while in forward motion.

Retreating-Blade Stall

In forward flight, the retreating blade tends to stall. Unlike fixed-wing airplanes that have their minimum speed limited by the stall, a helicopter's maximum speed is limited by blade stall. This is so because the retreating half of the disk must produce the same amount of lift as the advancing half. As the speed of the helicopter increases, the region of reverse flow (see Figure 8.16) increases in size and moves further from the rotor head. This requires more lift from the outer part of the retreating blade to compensate for the loss of lift. This is accomplished by increasing the angle of attack of the retreating blade, bringing it closer to the critical angle. Eventually, the tip stalls. On entry

> Over 3 million lives have been saved by helicopters in both peacetime and wartime operations since the first person was rescued from the sea in 1944.

into a retreating-blade stall, the pilot feels a vibration, the controls become sluggish, and the helicopter tends to tip to the side of the retreating blade. The causes for such a stall are typically high speed, high loading, low air density, and low rotor rpms.

The Power Curve

There are only two components to the power curve for a fixed-wing airplane: induced power and parasitic power. For a helicopter, there are three components, as shown in Figure 8.19. These components are induced power, parasitic power, and *profile power*. Profile power can be thought of as the parasitic power losses of the rotating blades. Induced power, as for an airplane, is the rate kinetic energy is given to the air to produce lift. We will start the discussion with a consideration of induced power.

> The helicopter is the only aircraft that has saved more lives than it has lost.

The induced power curve for a helicopter is more complicated than for an airplane. Remember, a wing flies into relatively still air. Nearly

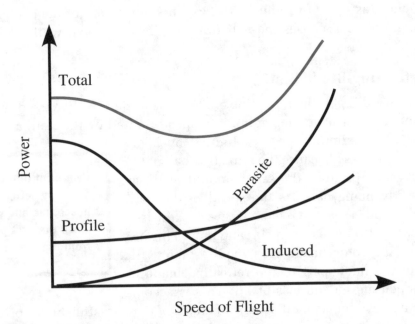

FIGURE 8.19   The three components of the power curve.

all the kinetic energy that is given to the air is the result of producing lift. The situation is not the same for a helicopter. The rotor blade is moving in the downwash of the blade ahead. To put it in simple terms, a helicopter that is hovering (zero forward speed) is flying in a downward flow of air, as is depicted in Figure 8.20. Thus it must use a great deal of power just to maintain altitude. One might look at this as like a person climbing a rope that is being lowered.

> Lieutenant Thomas Selfridge was the first casualty of a powered airplane crash. He was assigned by the U.S. Army to ride with Orville Wright. A cable snapped, breaking a propeller, leading to the crash.

As the helicopter starts to gain forward speed, the power necessary to maintain altitude decreases with increasing forward speed. This is so because part of the air being pumped through the disk is now coming from relatively still air. Figure 8.21 shows the airflow for a helicopter that is going forward at about 5 mi/h. The trailing half of the rotor disk has increased lift with increasing speed, and thus the forward tilt of the disk must be reduced as the speed increases.

By about 10 mi/h and above, the airflow through the disk is similar to that shown in Figure 8.22. This is also similar to the airflow over a

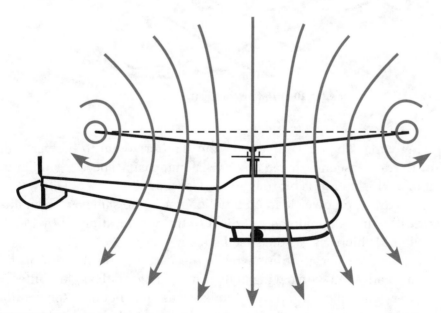

FIGURE 8.20   Airflow in hover.

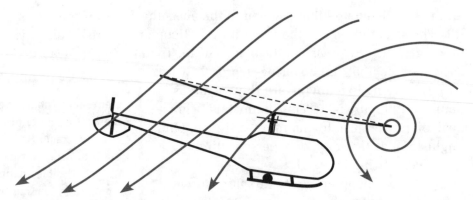

FIGURE 8.21 Airflow for 5 mi/h forward flight.

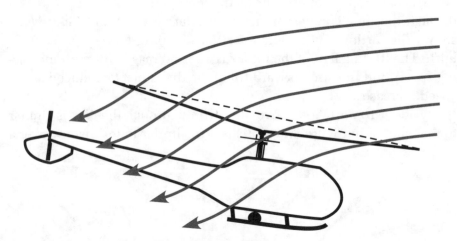

FIGURE 8.22 Airflow for 10 mi/h forward flight.

moving wing. The air is diverted from the horizontal to the vertical. Now the decrease in induced power with increasing speed is similar to that of a wing and falls off roughly as 1/speed.

The parasitic power load on a helicopter is the same as for an airplane. It is the rate that energy is lost to the surrounding air owing to collisions. It increases as the speed cubed.

As stated earlier, profile power is the parasitic power loss of the rotating blades. As with a wing, it does not change with angle of attack. However, since parasitic power increases as speed cubed, the profile power loss increases faster for the advancing blades than for the retreating blades. Thus profile power increases slowly as the helicopter's forward speed increases.

It should be noted that the literature on airplanes often has a great deal of material on drag and little on power. Just the opposite is true for helicopters. This is so because power is equal to drag times speed. Since the induced power is high at zero speed, this would imply an infinite drag. The solution is to play down drag for helicopters.

Because of the higher power required for a helicopter to hover (zero horizontal or vertical velocity), the hover ceiling is much lower than the service ceiling. For example, the service ceiling of the Blackhawk helicopter is 19,000 ft (5800 m). The hover ceiling, out of ground effect, is 7650 and 9365 feet (2330 and 2855 m) at 95 and 75°F (35 and 24°C), respectively.

Ground Effect

Most helicopters hover within *ground effect*. This is within a height above ground about equal to the rotor diameter. Thus, if the span of the rotor is 75 ft, then the helicopter is capable of hovering in ground effect up to about 75 ft. The importance of ground effect is that up to that limit, the air is physically compressed beneath the helicopter and the ground, and a cushion of the air is created. Because of ground effect, helicopters are able to hover anywhere from 5 to 80 ft above high mountain peaks, although they would not be able to hover at that altitude in midair. Thus mountains rescues can be made at great heights.

> The average infantryman in the South Pacific during World War II saw about 40 days of combat in 4 years. The average infantryman in Vietnam saw about 240 days of combat in 1 year, thanks to the mobility of the helicopter.

Ground effect is less affective for a helicopter than for a fixed-wing aircraft because induced power is a smaller fraction of the total power at low speeds. At hover, profile power makes a substantial contribution and is not affected by ground effect. Also, as for airplanes, the faster the helicopter is traveling across the ground, the less are the induced power requirements, and thus ground effect is less effective.

Running Takeoff

It sometimes happens that a helicopter is so heavily loaded that it can hover in ground effect but is not able to climb out of ground effect.

U.S. police and emergency rescue helicopters transport about 15,000 patients annually.

Thus it cannot take off vertically. If it is over smooth ground, it can make a *running takeoff*. The helicopter initially hovers just above the ground and then picks up forward speed. The faster it goes, the less power is required for flight. At some point, the required power is low enough to let the helicopter climb. This is why heavily loaded helicopters sometimes use runways like airplanes for takeoff.

Efficiency for Lift

As stated earlier, helicopters are inefficient aircraft. As an illustration, the specifications of a Cessna 172 and a Bell-47G helicopter are compared in Table 8.1.

TABLE 8.1   Comparison of the Specifications of a Cessna 172 and a Bell-47G Helicopter

	Cessna 172 Airplane	Bell-47G Helicopter
Maximum weight, lb (kg)	2300 (1045)	2347 (1067)
Useful load, lb (kg)	846 (398)	915 (416)
Power, hp	160	200
Fuel consumption, gal/h (liters/h)	7.5 (2 8.5)	23 (87)
Cruise speed, mi/h (km/h)	140 (224)	71 (113)
Range, mi (km)	725 (1160)	213 (341)
Rate of climb,* ft/min (m/min)	770 (233)	785 (238)
Service ceiling, ft (m)	14,000 (4,200)	8970 (2718)

*Sea level rate of climb.

As can be seen in the table, both aircraft have similar maximum weights and useful loads. The helicopter, however, consumes three times as much fuel and has half the cruise speed and only a third the range.

The major causes for the poor efficiency of the helicopter are that the rotors are relatively small in area. Thus they must produce lift with less air accelerated to a greater downward velocity than that for a fixed wing. The vertical velocity from the wing of the Cessna 172 is on the order of 10 mi/h (16 km/h) or less. The vertical velocity from the rotor of a helicopter can be as high as 60 to 100 mi/h (100 to 160 km/h). Since we know that the induced power is related to the vertical velocity of the air, the helicopter is at a great disadvantage. The induced power is proportional to the vertical velocity for the airplane, but the relationship is, of course, a little more complicated for a helicopter. This increased velocity of the helicopter is also due to the fact that the rotor blade is already in air with a downward velocity, and the lift is proportional to the change in the downward velocity caused by the rotor.

> There are 40,000 public parking spaces at the Dallas–Fort Worth airport. This is the most of any airport.

There are other causes for the inefficiency of the helicopter. In order to reduce the induced power requirements, the rotors are operated at high speeds at the expense of higher profile power requirements.

The fuselage is also in the downwash of the rotor, so some of the momentum transfer that would have produced lift is transferred to the fuselage. If the fuselage were to block all the downwash, there would be no lift but still a substantial consumption of power. This is one of the problems encountered by helicopters carrying large loads slung below them. The load blocks the downwash and reduces the available lift.

Finally, as already discussed, the tail rotor consumes power. The tail rotor can consume from 5 to 15 percent of the total power without producing lift or propulsion. A small helicopter with a 200-hp engine may require only 10 hp for the tail rotor. On the other hand, a 10,000-hp helicopter may require 1200 hp to compensate for the torque.

Autorotation

One of the requirements of a safe aircraft is that, on loss of power, it is able to descend to the ground safely in a controlled manner and make a safe landing. The power-off descent of a helicopter is called autorotation. In autorotation, the rotor continues to turn, creating lift and control. The energy is provided by the airflow through the bottom of the rotor disk. Here, energy comes from the conversion of potential energy into airflow to drive the rotor.

The helicopter has a free-wheeling clutch that allows the engine to drive the rotor but not the rotor to drive the engine. Thus, on loss of power, the rotor decouples from the engine and rotates freely. The tail rotor is coupled directly to the rotor, so it continues to turn, giving directional control.

The Concorde burns about 500 lb (225 kg) of fuel per passenger seat per hour. The Boeing 777 burns only about 40 lb (18 kg) per passenger seat per hour.

Autorotation is a hazardous procedure and thus usually is used only in emergency situations. The pilot has only 2 to 3 seconds to recognize the loss of power and make adjustments before too much energy has been lost from the rotor. Also, quick and precise actions must be taken at the moment of landing.

During powered flight, the airflow through the rotor disk is from the top and is directed downward (see Figure 8.23). In autorotation, the airflow is from underneath and is directed upward. In normal

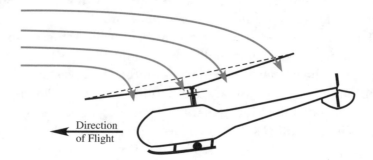

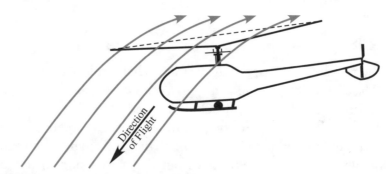

FIGURE 8.23 Airflow for normal flight and for autorotation.

flight, the engine overcomes the drag of the rotor in rotation. In autorotation, the relative wind drives the rotor at normal speed.

Vertical Autorotation

Most descents in autorotation are performed with forward velocity. For simplicity, we will start with a discussion of vertical autorotation in still air. In autorotation, the disk is divided into three regions, as shown in Figure 8.24. The regions are

> A maple seed is a one-bladed helicopter flying in autorotation.

1. *The driven region (also called the propeller region).* This is the outermost region of the rotor disk and normally consists of about 30 percent of the radius. The total aerodynamic force on the blade is inclined slightly backwards. This provides lift but creates a drag tending to slow down rotation of the rotor.

2. *The driving region or autorotative region.* This normally lies between 25 and 70 percent of the rotor radius. Here, the total aerodynamic force is tilted slightly forward, creating both lift and a forward force on the rotor.

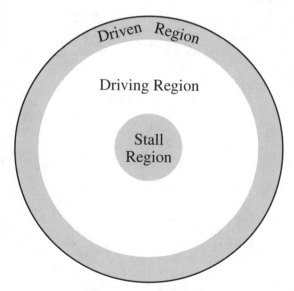

FIGURE 8.24   Three regions of disk in vertical autorotation.

3. *The stall region.* Normally is about 25 percent of the rotor closest to the rotor root. In this region, the rotor is stalling. Thus drag is produced without producing lift.

Let us now consider the details of the forces on the three regions of the rotor, referring to Figure 8.25. The upward flow of the air through the rotor disk combines with the rotational relative wind to produce a relative wind as seen by the rotor. This relative wind is both increased in speed over the air through the disk and strikes the rotor blades at a smaller angle, as shown by the inset in the figure. Since the speed of the rotor increases with distance from the hub, the relative wind is

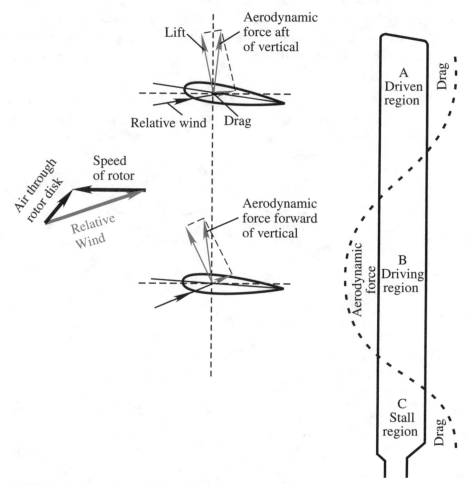

FIGURE 8.25   Details of the forces on the three regions of the rotor.

closest to horizontal at the rotor tip and steepest near the rotor root. Similar to an airplane's wing, the lift produced is perpendicular to the relative wind, and the drag is parallel to the relative wind. These two forces can be combined to give the total aerodynamic force. If the total aerodynamic force leans forward of the vertical, there is both lift and a forward force on the rotation or the rotor driving the rotation of the rotor. A backward-leaning total aerodynamic force yields a lift but also a drag that tries to slow down the rotation of the rotor.

Figure 8.25 shows the forces on the rotor in the driving region and driven region. The total aerodynamic force in the driven region is behind of the vertical. This region produces lift, but the net drag tends to decelerate the rotation of the rotor. The size of this region varies with the rotor pitch, rate of descent, and rotor rpms.

As one moves closer to the root of the rotor, the total aerodynamic force slowly rotates forward, going through an equilibrium point where it is vertical and then tilting forward in the driving region. Like the driven region, the size of the driving region varies with the rotor pitch, rate of descent, and rotor rpms. The pilot varies the size of the driving region with respect to the driven and stall regions by varying rotor pitch angle. This is done in order to adjust the autorotation rpms.

As one moves still further in toward the root of the rotor, the lift decreases because the rotor is approaching the stall condition by too high an angle of attack. At some point, the lift ends, and only the drag remains.

Forward Autorotation and Landing

Descent in forward autorotation is similar to vertical descent except that the altered airflow through the disk moves the driving and stall regions toward the retreating side of the rotor disk, where the angle of attack is higher. The lowest descent rate in autorotation is achieved in forward flight at the speed for the minimum power required for level flight. At this speed of forward flight, the descent rate is about half the descent rate in vertical autorotation. It should be noted that the lowest descent rate for a fixed-wing airplane is also at a forward speed corresponding to the minimum in the power curve.

The landing is made by using the kinetic energy in the rotor to slow the descent down. This is done by rapidly increasing the pitch on the rotor just above the ground. If the descent is too fast, there will not be sufficient energy in the blades to slow the descent speed adequately. Thus descents at very high airspeeds are more critical than those performed at the minimum-rate-of-descent airspeed.

Helicopter Height-Velocity Diagram

If the power is lost too close to the ground, a helicopter does not have enough altitude to enter autorotation. Therefore, there is a region at low speed and altitude at which the helicopter should not be operated. This region is shown in the height-velocity diagram in Figure 8.26, and its boundary is called *deadman's curve*. Above point A, at about 300 to 450 ft (100 to 150 m), there is sufficient altitude for the rotor to recover sufficient speed to make a safe landing. Below point *B*, 10 to 15 ft (3 to 5 m) above the ground, the helicopter will not have gained excessive speed before reaching the ground. At point *C*, with a sufficient speed of 20 to 30 mi/h (30 to 50 km/h), a safe landing can be made because of the reduction in autorotation descent rate with forward speed.

Because of the restrictions at low speeds in the height-velocity diagram, single-engine helicopters seldom make purely vertical takeoffs and landings. After a short vertical takeoff, the pilot will accelerate the craft forward to avoid the deadman's region.

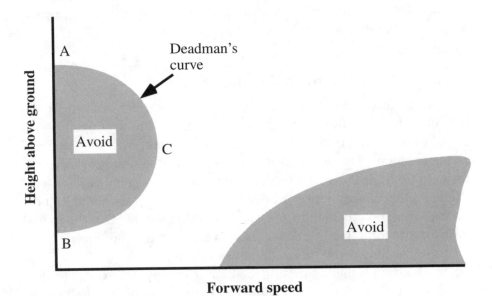

Figure 8.26   Height-velocity diagram.

There is also a restriction on high-speed flight near the ground shown in the figure. This is so because there is no time to reduce forward speed to avoid damage to the landing gear. This is particularly true for helicopters with skid-type landing gear.

Autogyros

Autogyros (also known as *autogiros*, *gyroplanes*, and *gyrocopters*) were the first practical rotary-wing aircraft. *Gyroplane* is the official term designation by the FAA, but autogyro is more common in the literature and will be used here. Today, all autogyros develop lift by a single multibladed rotor free spinning in autorotation. The rotor is tilted back at a small angle to catch the air. Since the rotor is unpowered, there is no torque from an engine to overcome, so no tail rotor is necessary. A rudder is used to make directional changes.

> Wind generators operate in autorotation.

Propulsion is achieved by a conventional propeller either pulling or pushing. An early design by the inventor of the autogyro, Juan de la Cierva of Spain, is shown in Figure 8.27. The autogyro in the figure is

FIGURE 8.27　Cierva C.19 autogyro.

the Cierva C.19. The attitude controls are similar to those of a fixed-wing aircraft with the ailerons on long rods.

Many of the key developments necessary for realization of the helicopter were first invented for the autogyro. These include the flapping hinge, the lead-lag hinge, and control by tilting the rotor disk by application of cyclic pitch. Before introduction of the cyclic, the attitude of the aircraft was controlled by use of a rudder and ailerons.

Autogyros have advantages and disadvantages over fixed-wing aircraft. Their primary advantage is that they are more stable in low-speed flight and cannot stall. When flown too slowly, they just descend slowly to the ground in a controlled manner. They also need less distance to take off and to land. Their major disadvantage is that they have high drag at high speeds. This makes the autogyro unsuitable for high speeds or long ranges.

The autogyro also has advantages and disadvantages over conventional helicopters. On the plus side, they are less complex, lighter, less expensive, and less complicated to operate. It is also less complicated to deal with the loss of an engine with an autogyro. Helicopters can be difficult to initiate autorotation, and often there is little time to do so. As mentioned, helicopters also have a deadman's area of flight where the helicopter is flying too low and slow to initiate autorotation in time. Autogyros have no such limitation. Finally, autogyros can fly faster than helicopters because the rotor is providing only lift, whereas the rotor of the helicopter also must provide thrust.

> Cierva came up with the idea of a hinged blade while watching a windmill with hinged blades at an opera.

In brief, the advantages of the autogyro are safety, ease of operation, and short takeoff and landing. Although modern autogyros can make "jump" takeoffs (discussed below) and land almost vertically, the clear advantage of the helicopter is that it can hover.

The first attempts at forward flight with an autogyro failed for two reasons. First, the asymmetry of lift owing to the advancing blades having more lift than the receding blades caused the craft to tilt to one side. Second, the gyroscopic effect of the rotor caused the craft to tilt on forward motion. Cierva conceived of the flapping hinge, which allowed the blades of the rotor to move up and down. This caused the blades with the higher lift to raise, reducing the angle of attack and balancing the lift on each side of the aircraft. These hinges also eliminated the gyroscopic effect.

Another problem that had to be solved was how to spin up the rotor before takeoff was attempted. There were several ways that spin-up was obtained. Small craft were spun up by hand. For larger craft, the spin-up was accomplished by an external power source such as an engine, horse, or a team of people. A third way was to deflect the backwash from the propeller through the rotor to get it spinning. Finally, a shaft and clutch arrangement was derived that could use the engine to prespin the rotor before takeoff.

An additional improvement on the autogyro was the invention, by Cierva, of the jump takeoff. The rotor was spun up to a speed greater than the minimum for a normal takeoff. When the power was removed from the rotor, the lack of torque caused the rotors to swing forward on the lead-lag hinges, increasing the angle of attack of the blades. The craft then would jump up into the air, and forward flight would be initiated by the propeller.

The Cierva C.30 is show in Figure 8.28. This is the first autogyro to make a jump takeoff as well as the first to use direct control by tilting

FIGURE 8.28 Cierva C.30 autogyro.

the rotor rather than by using a rudder and ailerons. The control bar, connected directly to the hub, can be seen in front of the pilot.

The autogyro never achieved much popularity. A year after development of the Cierva C.30 with its improved controls and jump takeoff, the first helicopter flew. This helicopter incorporated many of the innovations developed for the autogyro. Today, the autogyro is relegated to the world of ultralight aircraft, where it has been well received.

Wrapping It Up

Helicopters are much more complex in operation than conventional airplanes. Some people joke that they fly by beating the air into submission. Others claim that they don't fly at all but rather are so ugly that the earth repels them. Helicopters can be some of the most beautiful aircraft ever conceived. In fact, a Bell-47 (see Figure 8.1) hangs in the Museum of Modern Art in New York City. Helicopters have become engineering marvels that can bring hope to a person in distress or fear to an enemy on our modern battlefields.

> Inventor of the autogyro, Spaniard Juan de la Cierva, was killed in an airplane crash.

In the next and last chapter of this book we will give a brief introduction to the structure and materials of aircraft. This should give you a feel for some of the considerations that go into designing the airplane itself.

Structures

We have seen that the power required for lift goes as the load squared. Thus it is easy to see the incentive for a light structure. While we want a light structure, we also want it strong enough to carry the load without breaking.

Airplanes typically are designed to specific load requirements. For example, large transports are required to handle a 2.5g ultimate maneuver load. This means that the wing must be designed to carry two and a half times the gross weight of the airplane in any maneuver. A safety factor of 1.5 is used on top of this so that the airplane will not fail up to one and a half times that ultimate load. Acrobatic airplanes are required to have a design ultimate load of 6.

Designing an aircraft structure is a delicate balance of strength and weight. The object is a structure that meets all the structural load specifications in the regulations at the lightest weight possible. The first question that comes to many people's minds is, "Is the structure strong enough?"

> The Boeing B-314 Flying Boats were so large that there was a catwalk in the wings so that the mechanics could service the engines in flight.

A quick look at airplane crash histories shows that since the mid-1930s, structural failures in commercial aircraft are extremely rare. Today, even in light aircraft, in-flight structural failure is virtually nonexistent. However, aviation accident reports frequently report structural failure as a factor, but it is like reporting, "The car drove over the cliff, fell 200 ft, and on crashing into the ground, the vehicle experienced structural failure." On your next flight, then, as you see the wings bounce in turbulence, don't worry about the wings falling off; they won't.

Wings and Bridges

The simplest structure is a simple beam. Throw a log across a river to form a crude bridge, and you are using a simple beam. However, if the log does not reach the other side, you have created a simple *cantilevered* beam or, in this case, a diving board. Simple balsawood toy airplanes use a flat wing that is nothing more than a simple cantilevered beam. Thus, in the simplest sense, a wing is a cantilevered beam with one end fixed to the fuselage and the other end free. The normal load in flight is up, so the wingtip is bending up.

A collection of beams designed to carry load is called a *truss.* Trusses can be seen everywhere, if you know where to look. The "skeletons" of most modern buildings are trusses. Most bridges today are trusses. Figure 9.1 shows a classic railroad trestle that illustrates

FIGURE 9.1 Railroad trestle in British Columbia, Canada.

FIGURE 9.2   Truss structure of a wing.

the use of truss construction. Early airplanes used this concept in their structures. Figure 9.2 shows a wing of a classic airplane that is a truss.

In the earliest days of flight, it was believed that to reduce drag, wings had to be ultrathin. With little thickness, there was no way to build a strong enough truss structure that was fully enclosed within a single wing. Thus most airplanes were built as biplanes to create a boxlike truss structure. The upper and lower wings were strengthened by struts and diagonal wire bracing, as shown in Figure 9.3. This leads to a strong and efficient structure, but aerodynamically, the external bracing leads to very high drag.

Late in World War I, research in Germany led to thicker airfoils, which had no more drag than the thinner ones in use at the time. More of the structure could be embedded inside the wing, so fewer wires and struts were required. These wings were truss structures, cantilevered to the fuselage. Notable airplanes that pioneered this concept include the Fokker Dr.1 triplane, the Fokker D.VII, and the Fokker D.VIII. The triplane has taken on legendary status as the airplane Manfred von Richthofen, the "Red Baron," was flying when he was killed. It was really only an interim airplane that was not very successful. However, it was among the first airplanes to use

> The projected lifetime of a commercial airplane design is 70 years.

FIGURE 9.3 Biplane with wire bracing.

cantilevered wings. The prototype did not even have outboard struts, as shown in Figure 9.4. The Fokker D.VII was considered the Germans' best fighter in World War I. Note the lack of interplane wires on the Fokker D.VII in Figure 9.5, in contrast to the SPAD XIII shown in Figure 9.6. By the war's end, Fokker built a monoplane, the D.VIII, that used a fully cantilevered high-aspect-ratio wing. It had the elements of a modern aircraft.

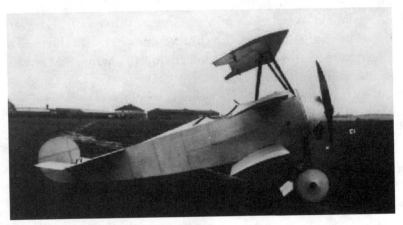

FIGURE 9.4 Fokker Dr.1 triplane prototype.

FIGURE 9.5 Fokker D.VII.

FIGURE 9.6 SPAD XIII.

The cotton used on early airplanes was attached to the wing ribs by a sturdy thread. The cotton literally was sewn onto the wing. In the early days, seamstresses were employed to do this work.

These early airplanes typically were made of a wood frame covered in cotton fabric. Wing loading was light, so the cotton was only required to hold 10 to 20 lb/ft² (50 to 100 N/m²). The load from the cotton was passed to wing ribs, trusses built in the shape of an airfoil, like those in Figure 9.2. The airfoils then passed the load to the spars, usually a front spar and rear spar. With this type of construction, lightweight and strong wings could be built. For smaller aircraft at lower speeds, this type of construction is still used today.

In 1917, when the United States entered World War I, a mass mobilization effort was undertaken to build airplanes for the war effort. Critical to these needs were raw materials for the allied cause. The best material for building wood and fabric airplanes is old-growth Sitka spruce. The old-growth spruce has a high strength-to-weight ratio; a long, straight grain; and few blemishes, such as knots. Sitka spruce was abundant in the Pacific Northwest of America. Some 30,000 young civilian men and soldiers were sent to southern Washington and northern Oregon coastal areas to log Sitka spruce. In the month of October 1918, over 22 million board feet of spruce boards were shipped to factories. It was this effort that caused the lumber industry in the Pacific Northwest to take off. Today, there is little old-growth Sitka spruce left.

It should be noted that many light airplanes today are built using the same type of construction, with aluminum skins to replace the cotton. Few airplanes still use fabric, and when they do, they use a synthetic Dacron or heavy nylon rather than cotton, which tends to rot after a few years. In the truss-type construction, the aluminum wing skins take no more load than the cotton they replaced, and aluminum is used only because of the longevity of the material.

It took over a decade after World War I before monoplanes with cantilevered wings became relatively common. At the same time, aluminum alloys that were much lighter than steel started to become available. Wing loadings became greater, and the wing skin started to be called on to take some of the bending load. The first *stressed-skin* airplanes were designed in the early 1930s. However, before stressed-skin airplanes, the forerunner of today's transports and fighters, a brief experiment in corrugated skins took place.

An efficient method of carrying load is to use corrugation. Bending against the axis of the corrugation is very stiff. The same weight of material can carry a greater load when corrugated. Cardboard boxes use corrugation internally to stiffen in one direction. It is fairly easy to bend corrugated cardboard along the corrugation but harder to bend against it. Several well-known airplanes used corrugated metals. The Junkers Ju-52 was the most widely used transport in Germany in the 1930s and is shown in Figures 9.7 and 2.33. In the late 1920s and early 1930s, Henry Ford experimented in aviation and built the Ford Tri-Motor, shown in Figure 9.8. The use of corrugation was limited because the corrugation

> The Fokker D.VII was the first airplane to exploit thick wings. As a result, it was the only airplane specifically mentioned in the Treaty of Versailles. The Germans were required to hand over every airplane to the Allies.

had to line up with the airflow, or drag would increase dramatically. On the wing, this is in the axis perpendicular to bending. Thus corru-

FIGURE 9.7 Junkers Ju-52.

FIGURE 9.8   Ford Tri-Motor with corrugated skin.

gation on the wing increases surface area and thus drag while doing little to stiffen the wing. In aircraft, corrugation was short-lived.

In the early 1930s, Boeing designed the 247, which many consider the "first modern airliner." It had retractable landing gear, landing flaps, and constant-speed propellers. It also used stressed skin, where the wing skins took a portion of the bending load. In an ideal structure, all components take the same load per cross-sectional area. Similar to pressure, the load per area is called *stress*. An optimized structure balances the stresses such that the stresses on all structural members are

FIGURE 9.9   Douglas DC-3.

equal. There are no structural members loafing while others are overworking. The Boeing 247 was among the first aircraft to successfully balance the loads in this way. This airplane led to the development of the Douglas DC-3, shown in Figure 9.9, which also used a stressed skin. Because of their sturdy construction, DC-3s still can be seen flying today, over 70 years after their introduction.

> Henry Ford tried to be a leader in airplane manufacturing. In 1927, the Ford Tri-Motor became the first successful commercial airplane.

The Wing Box

A modern transport is built such that the main structure of the wing is a big hollow beam. Figure 9.10 shows the stub of the *wing box* where the beam joins the fuselage of a modern airplane. The top and bottom of the beam are the upper and lower wing surfaces. The front and rear of the beam are the front and rear wing spars. The beam is reinforced with stiffeners on the interior of the skin running the length of the wing. Periodically, wing ribs are used perpendicular to

FIGURE 9.10 Wing box.

the wing-box axis to help with torsion, or twisting, loads. The wing box turns out to be an ideal place to store fuel. In the design of a modern jet transport, the amount of fuel that can fit inside the wing box is one of several constraints used to determine the size of the wing and the range of the aircraft.

> In 1991, the very first Boeing 727, the *Spirit of Seattle*, was retired after 64,492 flight hours, which is equivalent to 7.4 years in the air.

Modern jet wing construction is fundamentally different from the methods used prior to World War II and still in use on many light aircraft today. Instead of a light skin carrying the load to the ribs, which carry loads to one or two main spars, the wing box takes all the bending and torsion loads. The skin is now part of the beam. Instead of fabric or thin aluminum skins, modern transports use skins from 0.25 to 0.4 in (6 to 10 mm) thick. Add to these the stiffeners, and one has an extremely strong structure.

The understanding of material properties has become so sophisticated that different alloys of aluminum are used for the bottom and top wing skins. In normal flight, the bottom of the wing is in tension, whereas the top is in compression. This is illustrated in Figure 9.11. Therefore, the materials are optimized for their function of compression or tension. Moreover, the stiffness properties of the material also must be considered. The long-used B-52 bomber wingtip deflects up to 26 ft (8 m) from its drooping position on the ground to full deflection at a maximum-*g* pull-up. Figure 9.12 illustrates the B-52 on the ground, where the flexibility of the wing requires an outrigger wheel to prevent the tips from striking the ground. In the air, the upward bending of the outer tip is clear in Figure 9.13.

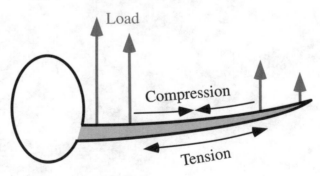

FIGURE 9.11 Wing bending under load.

FIGURE 9.12   B-52 taxiing with no load on the wing. (Photograph courtesy of the U.S. Air Force.)

FIGURE 9.13   B-52 in flight. (Photograph courtesy of the U.S. Air Force.)

Storing fuel in the wings has the added advantage that its weight counteracts the lifting load, alleviating some of the load the structure is required to take. Four-engine airplanes have the same advantage because the outboard engines hang from the wing and counter the lifting load.

The leading-edge and trailing-edge structures are attached to the front and aft spars. These include flaps, spoilers, and leading-edge devices. Note the flaps in Figure 9.12. There is a tradeoff on the size of the wing box and the size of the high-lift devices that attach to it. A larger wing box is stronger and carries more fuel but offers less room for flaps for a given wing area. A smaller wing box may help with increasing the high-lift capability but with reduced range.

Composites

Aluminum alloys have improved with each generation of aircraft. The technology of these alloys and the tools used to design with these alloys are reaching a limit. Just as we moved from steel tubing and fabric covering to aluminum structures, the time has come when one must look to new materials to continue the quest for lighter, stronger structures. These new materials are taking the form of carbon fibers, Kevlar, and carbon fiber–reinforced plastics (CFRPs).

Composite structures have been around for a long time. Technically, plywood is a composite material, using wood sheets and glue. The World War II British light bomber, the Mosquito, was built using plywood skin with a balsawood core. This made the Mosquito an extremely light and agile bomber. Fiberglass, another common composite, has been used on high-performance gliders for decades. In both military and transport aircraft, the use of composites has slowly crept in, first in secondary structures and later in primary structures.

Understanding Composites

To understand how composite materials work, we have to understand some basic properties of beams under load. In Figure 9.11, a winglike beam is experience a lifting load. The top surface is being compressed, whereas the bottom surface is being stretched.

Materials behave differently in tension and in compression. You can pull on two ends of a string, but you cannot push on them. You can push on a pile of loose bricks, but you cannot pull on them. Until steel became a practical building material, all large building structures had to be built such that every part was in compression. The available materials of stone and bricks, held together with mortar, were not strong in tension. Today, all concrete is poured over a matrix of rebar (reinforced bars made of iron) to strengthen the concrete in tension.

> Manfred Von Richthofen (the "Red Baron") flew the Fokker triplane, for which he is famous, for only 6 weeks and 19 of his 80 victories.

Composite materials are very much like the concrete and rebar combination. A resin, such as epoxy or other plastic polymer, is impregnated into fibers of graphite, aramid (Kevlar), or silicon glass. The fibers are extremely strong in tension, whereas the resin takes the compression load. Graphite (carbon) composites are stronger in tension than aluminum for a given weight but about the same in compression.

AN EXPERIMENT IN SHEAR

A simple way to understand how shear forces affect a structure is with an experiment using the inner cardboard roll from a paper towel or toilet paper roll. Take the roll between the palms of your two hands, and try to twist it. The cardboard roll is fairly stiff. Now, with a scissors, make a cut along the entire length of the tube. Now try to twist it again. The tube can no longer carry the torque. The reason is that the shear load cannot cross the slit you have made. Thus there is no strength in shear in the cut tube.

In a composite structure, one can build an "I-beam" of graphite-epoxy, fiberglass, or Kevlar, just as in metals. However, there is another option that is used commonly. A lightweight hard foam can be used to fill the volume between the top and bottom crosspieces. The foam is under much less stress because the load is spread out over a larger volume than the vertical piece of the I-beam. This type of structure is called a *sandwich* for obvious reasons.

Extending this idea from beams, large panels can be manufactured using composites or metals on the top and bottom surfaces with a hard foam or other lightweight core. Such applications for panels include commercial airplane floors. The panels are very light but strong and stiff. In fact, a big factor in the design of floor panels becomes the danger of puncture from women's high heels.

Fatigue

Commercial aircraft have very high utilization rates. An aircraft may be in operation 16 hours a day for an efficient airline. If the aircraft is used for short legs, it may experience 8 or 10 takeoffs and landings in a day. A takeoff and landing is called a *cycle*, and airplanes are designed to last for a certain number of cycles. During each cycle, the wing experiences a variation in loads that must be taken into account when designing the structure.

If you take a paper clip and bend it back and forth, it will break in about a dozen bends. With each takeoff and landing, the wing is bending from its ground-weighted position, full of fuel, to lifting the weight of the airplane in flight. This is exactly like the paper clip bending. Naturally, you would not want the wing to break after a dozen cycles.

For many parts on the airplane, it is not the ultimate load that sizes the part, but the fatigue life. If the part must last 30,000 to 40,000 cycles, it must be very durable.

In addition to designing for fatigue life, aircraft structures are inspected continually. All general-aviation aircraft are required to have an annual inspection, in which a certified mechanic must inspect all critical parts of the structure to make sure that there are no fatigue cracks or corrosion. Commercial aircraft have inspection schedules, the most frequent occurring every 100 hours of flight and the most extensive, called a *D check*, that might happen every few years. In a D check, the aircraft is stripped down to the bare structure for detailed inspection. This can take up to a month.

Wrapping It Up

An ideal aircraft structure would be designed so that every part fails at exactly the same limit load and fatigues at exactly the same number of cycles. And these failure conditions are selected so that they just cannot happen under normal operating conditions. The ideal structure also would have no margin above these conditions because that just means extra weight.

The reality of aircraft structures is that they are usually overdesigned. No aircraft can be certified if it does not meet the requirements, but there is no regulatory penalty for exceeding them, only a performance penalty. With the expense required to design a large aircraft, the larger Boeing and Airbus transports usually are designed much closer to tolerances. Smaller general-aviation airplanes typically are overbuilt, meaning that many individual parts are stronger and heavier than they need to be. This does not mean that the airplane as a whole has excess margin, but many individual parts do.

Gustave Eiffel is considered the father of applied aerodynamics. In his later years, he built wind tunnels and designed airfoils. Of course, he is best known for designing the Eiffel Tower.

The primary objective of the aircraft structure is to carry the required flight loads with as little weight as possible. This has led from wood and fabric airplanes to stressed-skin composites. Today's airplanes use the most advanced lightweight materials and the most advanced structural design and analysis tools to produce the most efficient structures possible.

Basic Concepts

A serious discussion of aeronautics requires a basic set of concepts and terminology. Frequently, in the process of learning, the language becomes an unrecognized barrier to the uninitiated. We present here some basic terminology and concepts we hope will alleviate some of this. The material has been put in an appendix rather in the first chapter, as in the first edition of this book, to make the book more readable for those who are already familiar with most of the concepts and structures of an airplane.

Airplane Nomenclature

Some readers may be familiar with the language of airplanes and others not. We encourage all to read the following sections of this appendix to ensure that the book is easily understood. Those familiar with the major parts of an airplane, the operation of the control surfaces, and the basic operation of an airplane can skip ahead to the section on kinetic energy.

The Airplane

Figure A.1 shows the main components of a high-winged airplane. The *airframe* consists of the *fuselage*, which is the main component of the airplane, the *wings*, and the *empennage*. The empennage (sometimes called the "tail feathers") is the tail assembly consisting of the *horizontal stabilizer*, the *elevators*, the *vertical stabilizer*, and the *rudder*. The elevators are used to adjust, or control, the *pitch* (nose up/down attitude) of the airplane. The elevators are connected to the control wheel or stick of the airplane and are moved by the forward and backward motion of the control wheel. On some airplanes, the entire horizontal stabilizer is the elevator, as shown in Figure A.2. This is called a *stabilator*. The rudder is used to make small directional changes and turns. Two pedals on the floor operate the rudder, used to assist directional control.

Most airplanes have small hinged sections on the trailing edges of the elevators and sometimes on the rudder called *trim tabs*, as shown in Figure A.2. These tabs move in the opposite direction to the control surface. The purpose of the trim tabs is to reduce the necessary force on the control wheel, also called a *yoke*, for the pilot to maintain a desired flight attitude.

Most modern airplanes have single wings mounted either above or below the fuselage. Most, but not all, high-winged airplanes have wings that are supported by *struts*. Struts allow for a lighter wing but at the expense of a little more drag (resistance to motion through the air).

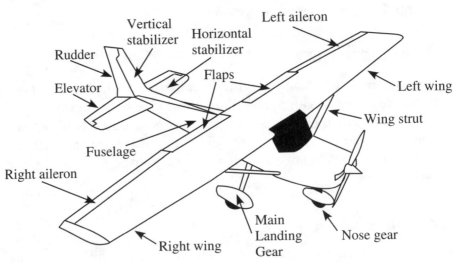

FIGURE A.1 Main components of an airplane.

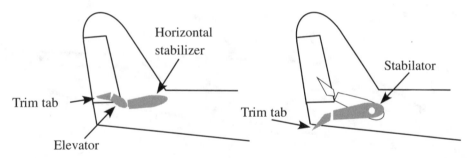

FIGURE A.2 **The empennage.**

The movable surfaces on the outer trailing edges of the wings are the *ailerons*, which are used for *roll control* (rotation around the center axis of the fuselage). They are operated by rotation of the control wheel or by the left-right movement of the stick. The ailerons are coupled so that when one swings up, the other swings down. Control surfaces will be discussed in detail below.

The hinged portions on the inboard part of the trailing edges of the wings are the *flaps*. These are used to produce greater lift at low speeds and to provide increased drag on landing. This increased drag helps to reduce the speed of the airplane and to steepen the landing approach angle. Flaps are discussed in detail in Chapter 2.

Small airplanes have two configurations of landing gear. *Tricycle landing gear* has the *main landing gear* just behind the center of balance of the airplane and a steerable *nose gear* up forward. The *tail dragger* has the main landing gear forward of the center of balance and a small steerable wheel at the tail. The nose gear and the *tail wheel* are steered with the rudder pedals.

Airfoils and Wings

The *airfoil* is the cross-sectional shape of the wing. As shown in Figure A.3, an airfoil is the shape seen in a slice of a wing. Besides the wing, propellers and the tail surfaces are also airfoils. Even aeronautical engineers (in discussions) sometimes use the terms *wing* and *airfoil* interchangeably. However, an airfoil is just the shape seen in a slice of the wing and not the wing itself. For some wings, slices taken at different places along their length will reveal different airfoils.

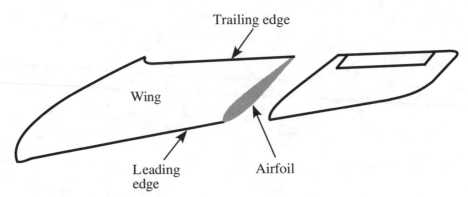

FIGURE A.3 **The wing and airfoil.**

An airfoil, as shown in Figure A.3, has a *leading edge* and a *trailing edge*. As detailed in Figure A.4, a *chord* and a *camber* also character-ize an airfoil. The chord is an imaginary straight line connecting the leading edge with the trailing edge. The chord is used for determining the *geometric angle of attack* (discussed below) and for determining the area of a wing.

The *mean camber line* is a line equidistance from the upper and lower surfaces of the wing. The camber is the curvature of the mean camber line. A wing that has an airfoil with a great deal of curvature in its mean camber line is said to be a *highly cambered wing*. A symmet-ric airfoil has no camber.

An airfoil with lift also has an angle of attack, as shown in the fig-ure. The *relative wind* is the direction of the wind at some distance from the wing. It is parallel to the motion of the wing. The velocity of the relative wind is equal to the speed of the wing. In aeronautics, the *geometric angle of attack* is defined as the angle between the *mean chord* of the airfoil and the direction of the relative wind.

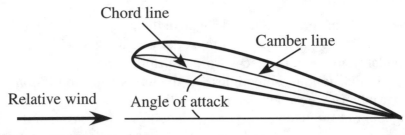

FIGURE A.4 **Airfoil nomenclature.**

A useful measure of a wing is its *aspect ratio*. The aspect ratio is defined as the wing's *span* divided by the average or *mean chord length*. The span is the length of the wing measured from wingtip to wingtip. The mean chord length is the average chord length along the wing. The area of the wing is just the span times the mean chord length. Most wings on small general-aviation airplanes have aspect ratios of about 6 to 8. This means that the wing is 6 to 8 times longer than its average width.

Axes of Control

An airplane moves in three dimensions called *roll*, *pitch*, and *yaw*, as illustrated in Figure A.5. *Roll* is rotation about the longitudinal axis that goes down the center of the fuselage. The ailerons control rotation about the roll axis. Pitch is rotation about the lateral axis of rotation, which is an axis parallel to the long dimension of the wings. The elevators control the pitch of the airplane. By controlling the pitch of the airplane, the elevators also control the angle of attack of the wing. To

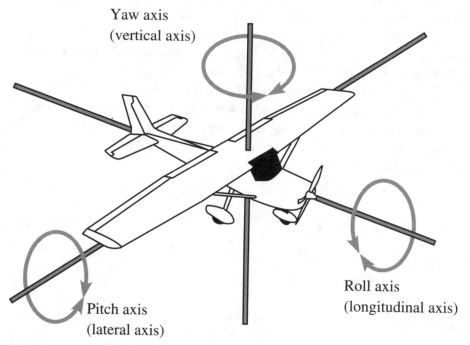

FIGURE A.5   Axis of rotation of an airplane.

increase the angle of attack, the entire airplane is rotated up. As we will see, this control or the angle of attack is the key to adjustment of the lift of the wings. Finally, *yaw*, which is controlled by the rudder, is rotation about the vertical axis, which is a line that goes vertically through the center of the wing. It is important to note that all three axes go through the *center of gravity* (often abbreviated *cg*) of the airplane. The center of gravity is the balance point of the airplane. Or equivalently, all the weight of the airplane can be considered to be at that one point.

The Turn

One common misconception by those who are not pilots is that, as with a boat, the rudder is the control used for making a turn. Although very small direction changes can be made with the rudder, the ailerons are use in making turns. The airplane is rolled to an angle in the direction of the desired turn. The lift developed is perpendicular to the top of the wing. In straight-and-level flight, this is straight up. As shown in Figure A.6, when the airplane rolls to some angle, the direction of lift is now at an angle, with part of the lift force used for turning and part used to support the weight of the airplane. In a turn, the rudder is used only to make small corrections and *coordinate* the turn.

As shown in the figure, the pilot feels a force equal to the lift, but in the opposite direction. Occasionally, in the book, we referred to the *2g turn*. A 2g turn is a turn where the force felt by the pilot is twice the force of gravity (2g) and the force, or *load*, on the wing has been doubled. In aeronautical terms, the *load factor*, which is the load divided by the weight, on the airplane in a 2g turn is 2. Figure A.7 shows the load fac-

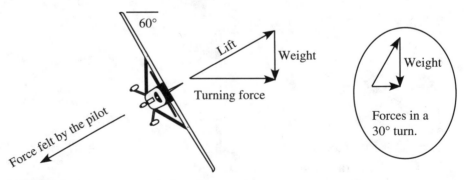

FIGURE A.6 Forces on an airplane in a turn.

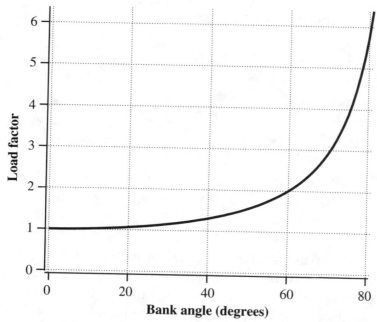

FIGURE A.7 Load factor as a function of bank angle.

tor as a function of *bank angle* for any airplane in flight. One thing to understand is that the forces on the pilot (or the load) are only related to the bank angle, which is the angle made by the wing and the horizon. In Figure A.6, the bank angle is 60 degrees. The vertical part of the lift always must be equal to the weight of the airplane if the altitude of the airplane does not change during the turn. This is called a *level turn*. Thus the steeper the bank angle, the greater is the lift, and therefore the greater is the force felt by the pilot. The inset in the figure shows the forces of a 30-degree turn for comparison. The weight part of the lift is the same, but the other two forces are less. A 2g turn is achieved by banking the airplane at an angle of 60 degrees, independent of the speed of the airplane. Turns are discussed in more detail in Chapter 6.

The Four Forces

There are four forces associated with the flight of an airplane. These forces, illustrated in Figure A.8, are lift, weight, thrust, and drag. In *straight-and-level flight* (not changing speed, direction, or altitude),

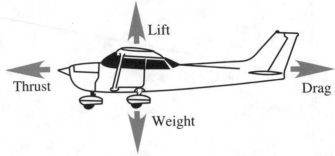

Straight-and-level flight

Lift

Thrust

Drag

Weight

FIGURE A.8   The four forces on an airplane in straight-and-level flight.

the net lift on the airplane is equal to its weight. We say *net lift* because, for a conventional airplane design, the horizontal stabilizer pulls down, putting an additional load on the wings. The thrust produced by the engine is equal to the drag, which is caused by air friction and the work done to produce the lift.

Mach Number

One important parameter in describing high-speed flight is the *Mach number*. The Mach number is simply the speed of the plane, or speed of the air, measured in units of the speed of sound. Thus a plane traveling at a speed of Mach 2 is going twice the speed of sound. The speed of sound is fundamental for flight because it is the speed of communication between the airplane and the air and between one part of the air and another. In Chapter 5, it is shown that as the speed of an airplane approaches Mach 1, there are dramatic changes in the airplane's performance. One change that affects performance is that the air ceases to separate before the wing arrives and instead collides energetically with the wing.

The speed of sound is not a constant in air. In particular, it changes with air temperature and thus altitude. As the air temperature decreases with altitude, so does the speed of sound, although not as quickly. At sea level, the value of Mach 1 is about 760 mi/h (1220 km/h). The speed of sound decreases with altitude to about 35,000 ft (about 11,000 m), where the value is about 660 mi/h (1060 km/h). The speed of sound then remains essentially constant to an altitude of 80,000 ft (24,000 m). No airplanes fly above this altitude, with the exception of the Space Shuttle on its way back from space.

Kinetic Energy

Kinetic energy is the energy of an object because it is moving. The energy difference between a bullet sitting on a table and one flying through the air is the kinetic energy. To be technical, if the bullet had a mass m (say, in grams, for example) and were moving at a velocity v, its kinetic energy would be $\frac{1}{2}mv^2$. (There, we have reached the highest level of math complexity that one needs to understand flight.)

Because the book discusses the movement of air and the production of propulsion by the acceleration of air or exhaust, it is important that when we say *kinetic energy*, it is understood that we mean the "energy owing to motion." It is that simple.

Air Pressures

For those reading this appendix first, we should spend just a little time discussing air pressure to remove some common misunderstandings. The following discussion is only strictly true for air moving at speeds below about Mach 0.3 (three-tenths the speed of sound), where the air can be considered incompressible. This is discussed in greater detail in Chapter 1.

Many of us have seen pictures of air passing through a tube that narrows as in Figure A.9. The figure will be referred to often in text that says something like, "As the area of the tube narrows, the flow velocity must increase. If no other force acts on the fluid, the pressure at point A must be greater than the pressure at point B." This is the Bernoulli relationship that some of you are familiar with in the explanation of lift in flight. At first, the meaning of "the pressure at point A" seems obvious. What is never said in physics books is that the pressure referred to is measured perpendicular to the direction of flow. It is also not said that there are two other pressures associated with the air at point A. One of them has increased, and the other has remained the same. Aeronautical engineers understand this concept, but somehow the information has not made it to the aviation community.

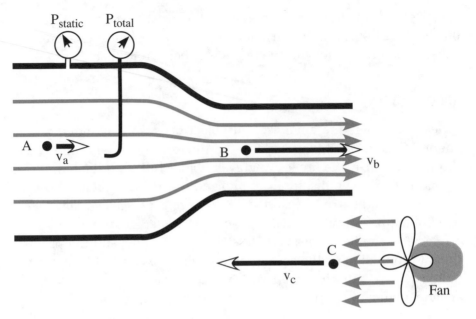

FIGURE A.9 Airflow and pressures.

The first of the three pressures associated with flowing air is the *total pressure.* This is measured by bringing the flowing air to a stop. In Figure A.9, this is measured by placing a tube facing into the airflow. The air stops in the tube, and the total pressure is measured (P_{total} in the figure). In the situation in the figure, P_{total} is the same at both points A and B. In the language of pilots, this is also known as the *pitot pressure,* and the figure illustrates a *pitot tube.*

The second pressure to consider is the *static pressure* (P_{static}), which is measured perpendicular to the airflow through a hole in the wall. This is the pressure most often referred to when the air pressure is discussed in aerodynamics. In the figure, the static pressure is higher at point A than at point B.

The third pressure is the *dynamic pressure* ($P_{dynamic}$), which is the pressure owing to the motion of the air and is a pressure parallel to the flow of air. The dynamic pressure is proportional to the kinetic energy in the air. Thus the faster the air goes, the higher is the dynamic pressure. This may seem a little complicated, so let us try to put it all together.

The total pressure P_{total} is the sum of the static and dynamic pressures ($P_{static} + P_{dynamic}$). We have shown how to measure the total pressure and the static pressure. How is the dynamic pressure of the air measured? Look at the setup in Figure A.10. Between the tube that measures the total pressure and the tube that measures the static pressure, there is placed a *differential pressure gauge*. This is a gauge that measures the difference in pressure between the two ports, which is the difference between total and static pressure. Since static plus dynamic pressure is equal to total pressure, this difference between total and static pressures is the dynamic pressure.

If no energy is added to the air (by some mechanism such as a propeller), the total pressure remains the same, and an increase in dynamic pressure causes a decrease in static pressure. Thus, when one reads that the pressure of the air decreases because it is going faster, the pressure referred to is the static pressure. However, what if energy is added to the air? In the right-hand corner of Figure A.9 is a picture of a fan. What has happened to the air pressures at point C? The fan is accelerating the air, and thus doing work on the air. Therefore, the dynamic pressure has increased. Since the air is not confined, the static pressure is the same as the surrounding environment and has not changed. Thus the total pressure has increased.

The Pitot Tube

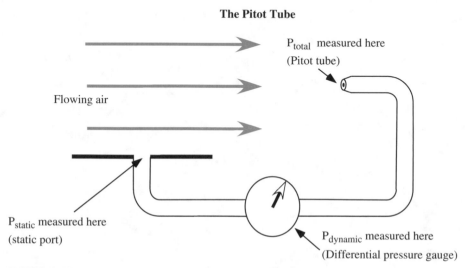

FIGURE A.10   Measurement of pressures with a pitot tube.

The moral of this story is that when someone refers to the air pressure of moving air, he or she is probably referring to static pressure (although he or she may not know it). It is also wrong to think that just because air is flowing faster, the (static) pressure has decreased. This topic is covered in more detail in Appendix B.

The Pitot Tube

As mentioned earlier, the tube measuring the total pressure in Figure A.10 is called a *pitot tube*, which, along with static pressure, is used for measuring the airspeed of an airplane. Several pitot tubes can be seen on the front of large jets at the airport and can be seen in Figure 6.4. A single pitot tube can be seen protruding (or hanging) from the wing of a small airplane. The hole that measures the static pressure of the air is the *static port*. This port is usually on the side of the airplane fuselage near the front, although it is also occasionally placed on the side of the pitot tube itself. In an airplane, there is a gauge that measures the difference between these two devices and is calibrated in speed. This is the *airspeed indicator*. As will be discussed in Chapter 6, this measurement is of *indicated air speed*, which must be corrected for air density to give *true airspeed*.

Misapplications of Bernoulli's Principle

Most descriptions of lift that involve Bernoulli's principle depend on the shape of the wing. But we show in this book that the shape of the wing affects the efficiency and stall characteristic of a wing but not the lift. These descriptions also start out with the acceleration of the air causing a lowering of the pressure through Bernoulli's principle. However, it is the lowering of the pressure over the wing that accelerates the air. This reversal of cause and effect flies in the face of Newton's first law by suggesting that the acceleration leads to a force.

Finally, some people will argue that Bernoulli's principle is simply the conservation of energy and so must be applicable. However, for Bernoulli to be applicable, there can be no *external work* done, and the air must be in *equilibrium*. That is, no energy can be added to or removed from the airflow. Before the wing came by, the air was standing still. Afterward, a great deal of air has been accelerated down (opposite to the direction of the lift). As we saw, this is the source of induced power and induced drag. A great deal of energy has been added to the air. Thus either we must have an engine to oppose this drag, which is a source of *external work*, or the wing will decelerate, which means it is not in *equilibrium*. Thus Bernoulli's principle is not applicable to a real, three-dimensional wing.

> We wish to point out that this may appear to contradict an accepted application of Bernoulli. However, what is suggested is that Bernoulli is not applicable unless the wing is 100 percent efficient. That is, there is no energy loss in producing the lift. In classical presentations of lift on a two-dimensional airfoil, the wing is 100 percent efficient, and so Bernoulli can apply. This is so because an infinite amount of air is accelerated at almost zero velocity. It is only when considering a true, three-dimensional wing, which cannot be 100 percent efficient, that Bernoulli becomes inapplicable. The pressures and velocities of the air passing over a real wing producing lift are not related by Bernoulli's principle. In this case, a great deal of energy is given to the air, so either energy has to be added (with an engine) or there can be no lift.

Demonstrations of Bernoulli's principle are often given as explanations of the physics of lift. They are truly demonstrations of lift, but certainly not of Bernoulli's principle. We are often taught only part of what is necessary to understand the applications of Bernoulli's equation. This has been the source of a great many misconceptions that have been propagated enthusiastically. When we are first introduced to Bernoulli's equation, it is always in respect to a fluid flowing in a pipe with a restriction. Since mass must be conserved, the flow through the smaller cross section is faster. Since energy is conserved, the faster-flowing fluid has a lower pressure. This is often all that we are taught. And from this, most of us have come away with the belief that if air is moving faster, it has a lower pressure. On giving it deeper thought, we might rightfully assume that the lowered pressure is measured perpendicular to the flow (the static pressure) because we know that if we put our hand in the path of this faster-flowing air, we would feel an increase in pressure.

Because of this point of view, some very interesting things are taught. First, take the example of the Bernoulli strip illustrated in Figure B.1. The Bernoulli strip is a narrow piece of paper across the top of which one blows to produce lift. It rises into the airstream and is clearly an example of lift on a wing. "The air goes faster over the top; thus the pressure is lower, and the paper rises." Or at least so goes the explanation.

Another common example of the physics of lift is that of a Ping-Pong ball supported by a vertical jet of air (Figure B.2a). The argument

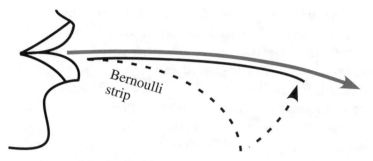

FIGURE B.1 The Bernoulli strip.

is that since the air is moving, the static pressure is lower. When the ball moves to the side, it comes into contact with the still air that is of a higher pressure. The ball is pushed back into the flow.

Before we explain what is missing in our understanding of Bernoulli's equation, let us revisit the static port on an airplane. An airplane has a small port somewhere on the side of its fuselage where

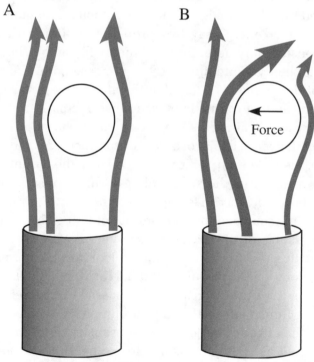

FIGURE B.2 Ping-Pong ball in a jet of air.

the static pressure is measured by the instruments, such as the altimeter. This port provides a very accurate static pressure measurement, even though air is passing over it at a high speed. If one watches the altimeter when the engine is started and the propeller blows air across the static port, the indicated altitude does not change. However, the altimeter gives a very sensitive measure of pressure. So what is wrong with our understanding of the Bernoulli principle?

In aeronautics, Bernoulli's equation is well understood. Ignoring the change in altitude and compressibility of the airflow, one can write Bernoulli's equation as

$$P_{static} + \tfrac{1}{2}\rho v_2 = P_{total}$$

Here, P_{static} is the static pressure measured perpendicular to the flow. The second term is referred to as the *dynamic pressure*, where the density of the air is ρ and v is its speed. Thus the dynamic pressure is a measure of the kinetic energy of the air. P_{total} is the total pressure.

In a confined pipe, the sum of the dynamic pressure and the static pressure is a constant. If we know that constant, the static pressure can be calculated from knowledge of the air's velocity. The same is not true for air that is given energy, say, by a propeller or our breath. We do not know the constant anymore. In fact, kinetic energy has been given to the air. Thus the dynamic pressure has increased, but the static pressure has not decreased. The fact that the air is moving faster does not necessarily mean that the static pressure has decreased.

Let us now look again at the Ping-Pong ball in the jet of air. First, one might reason that since the jet of air is not confined, if it had a lower static pressure, the surrounding air would collapse the jet until it had the same static pressure as the surroundings. This is reasonable because there would be a difference in (static) pressures and no barrier to separate them. In fact, the source of the jet of air has only increased the dynamic and total pressures of the air. Likewise, one's breath does not have a decreased static pressure. Thus one must look for another explanation for the Ping-Pong ball in a jet of air and the lifting of the Bernoulli strip.

The answer lies in Newton's laws (discussed in Chapter 1). Remember, a flowing fluid such as air wraps around a solid object. When the ball is near the edge of the jet of air, there is asymmetric flow of air around the ball, as in Figure B.2b, and momentum transfer causes a force to push the ball back in, just like the lift on a wing.

The same is true with the Bernoulli strip. The air bends over the paper strip. Newton's first law says that this requires a force on the air. Newton's third law says that an equal and opposite force is exerted on the paper. The paper is lifted. Our incomplete understanding of its application causes most of the problems with applications of Bernoulli's principle. We have been led to assume that if air is flowing, its static pressure has been lowered. This, of course, is not necessarily so.

There are two other phenomena often attributed to Bernoulli's principle. The first is the situation where one blows between two Ping-Pong balls hanging on strings, as shown in Figure B.3. The result is that they swing in toward each other. Here, we just have the same phenomenon as the Ping-Pong ball in the jet of air. In this case, though, there are two balls instead of one.

A more interesting misapplication of Bernoulli's principle is in the explanation of the curved flight of a spinning baseball. Let us start the discussion by examining the airflow around a nonspinning ball in flight, as shown in Figure B.4a. In the figure, the ball is traveling from right to left. The air splits evenly around the ball, and the only force on the ball is the drag of the air.

Figure B.4b shows the effect on the air when the ball is spinning. Here we have removed the effect of the ball traveling through the air. The roughness of the ball causes air to be dragged around, forming a boundary layer moving in the direction of rotation. The stitching on the ball enhances this boundary-layer formation. The surface is sometimes illegally roughened by the pitcher to enhance the effect.

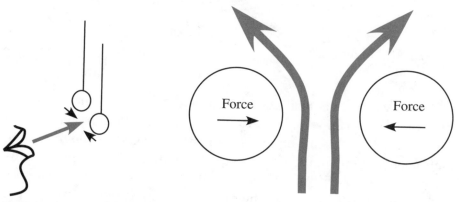

FIGURE B.3   Two Ping-Pong balls on strings.

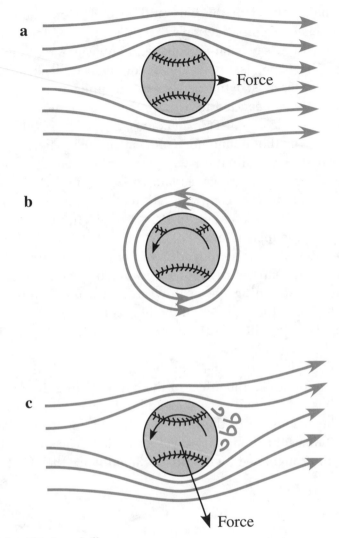

FIGURE B.4 The curve ball.

When we put the airflows of Figures B.4a and B.4b together, we find that the airflow looks very much like that over a wing, only upside down in this example. The air that travels over the top of the ball meets the oncoming flow around the ball and loses energy. This causes the air to separate from the ball fairly early. The air that goes under the ball is traveling in the same direction as the air around the ball. This air stays attached longer. The result is a net upwash behind the ball, and thus there is a downward force on the ball. Therefore, a spinning ball will feel a force perpendicular to its direction of travel. The backspin given to a dimpled golf ball causes it to experience a lifting force in the same way.

There are other examples of the misapplication of Bernoulli's principle, but we hope you get the idea. The next time you hear Bernoulli given credit for some phenomenon, think it through and see if you really believe what you are being told.

INDEX